Sulphuric Utopias

Inside Technology

Edited by Wiebe E. Bijker, W. Bernard Carlson, and Trevor J. Pinch

A series list appears at the back of the book.

Sulphuric Utopias

A History of Maritime Fumigation

Lukas Engelmann and Christos Lynteris

The MIT Press
Cambridge, Massachusetts
London, England

The open access edition of this book was made possible by generous funding from Arcadia—a charitable fund of Lisbet Rausing and Peter Baldwin.

This book was set in ITC Stone Serif Std and ITC Stone Sans Std by Toppan Best-set Premedia Limited.

Library of Congress Cataloging-in-Publication Data

Names: Engelmann, Lukas, 1981- author. | Lynteris, Christos, author.
Title: Sulphuric utopias : a history of maritime fumigation / Lukas
    Engelmann and Christos Lynteris.
Description: Cambridge, Massachusetts : The MIT Press, [2019] | Series:
    Inside technology | Includes bibliographical references and index.
Identifiers: LCCN 2019029666 | ISBN 9780262538732 (paperback)
Subjects: LCSH: Ships--Fumigation--History--20th century. |
    Ships--Disinfection--History--20th century. | Chemical apparatus.
Classification: LCC VM483 .E64 2019 | DDC 628.9/6--dc23
LC record available at https://lccn.loc.gov/2019029666

Contents

List of Figures

Preface

This book is the result of extensive interdisciplinary scholarship and collaboration. Research leading to *Sulphuric Utopias* brings together the history of science with medical anthropology and science and technology studies (STS), integrating in the process aspects of colonial, world, and oceanographic history and the history of infectious diseases. We have taken recourse to this ensemble of perspectives so as to write the story of a technological contraption, the Clayton apparatus, as a history of sulphuric utopias. This is a history of a set of concepts, practices, and technologies that revolve around the idea of fumigation. At the turn of the nineteenth century and until the 1930s, filling the holds of merchant vessels with gaseous substances that acted on trapped air, as well as on the surfaces of walls and the cargo itself enabled a utopian vision of a maritime space free from disease. Sulphur (we maintain the British spelling of the word, as it was used on both sides of the Atlantic at the time), or rather sulphuric acid gas in particular, was imagined to act with chemical indifference on any disease agent, as well as on the minds of people, thus addressing a range of concepts of disease transmission while providing a spectacular hygienic intervention.

With this book we then address an almost classic history of modernization. Medical officers at the end of the nineteenth century reshaped and rearticulated existing practices in the name of sound scientific principles. They utilized the age-old alchemy of sulphur and ancient traditions of smoking-out and fumigation. They experimented, measured, and tested with the aim of defining the exact capacity of different gases against bacteria, insects, and rodents. But, as we argue in this book, this process of rationalization did not shed its premodern, and perhaps philosophical associations, but instead mobilized them so as to install a specter of total hygienic environments. To overcome the costly, impractical, and untrustworthy system of involuntary detention in quarantine, it was necessary to continue to act on the imaginations of people as much as it required acting on pathogens and disease vectors. As such, this book presents the modern history of the desire to return a built environment, affected and infected by disease, into a status of hygienic purity. It is the history of how the old miasmatic imaginary about the omnipresent effluvia of disease was modernized through a chemically based sanitation process. Fumigation with sulphuric acid gas, on the one hand, acted upon bacteriologically specific agents, insects, and rodents, while, on the other hand, it maintained an unspecific hygienic promise: a *sulphuric utopia*.

In some ways, this is a history of a failed technology, which populated only a comparably short slice of the modern history of maritime sanitation. Globally dispersed archival sources and a widely forgotten history of fumigation with sulphuric acid gas have contributed to the astonishing invisibility of this story in the history of global health, the history of capitalism, and the history of maritime trade. The Clayton machine appears at best as a footnote to these histories, and its description is often incomplete and inconsistent.

The global dimensions and lasting impact of this effort of establishing a barrier against epidemic diseases are largely overlooked. And so, it is no accident that we encountered the global presence of its principal apparatus, the Clayton, through research on the history of the third plague pandemic (1894–1959), carried out in the European Research Council (ERC) funded project "Visual Representations of the Third Plague Pandemic," led by Dr. Christos Lynteris (grant agreement no. 336564). It was in the global archive of this plague pandemic, following its crossing routes across continents, empires, and epistemic frameworks, where the dimensions and the historical significance of a new system of maritime sanitation became first visible to us. And it was in this global pandemic—the first to be understood and acted as such—where we traced the steps of the globalization of an early technoscientific object, which was installed to overcome the past of unspecific diseases and to enable the uninterrupted flow of goods and people in the global distribution of capitalism.

Acknowledgments

Research leading to this book was funded by a European Research Council Starting Grant (under the European Union's Seventh Framework Programme/ERC grant agreement no 336564) (PI Christos Lynteris), a Fondation Brocher Residency Fellowship (for Lukas Engelmann and Christos Lynteris, Spring 2017), research funding from Cambridge Humanities Research Grants for archival research in New Orleans (for Lukas Engelmann, Fall 2017) and a Rockefeller Archive Center Research Stipend (for Lukas Engelmann, July 2017). We would like to thank the ERC, the Fondation Brocher, the Rockefeller Archive, and Cambridge Humanities Research Grants for their generous support in this research. We would also like to express our gratitude to our colleagues at the Centre for Research in the Arts, Social Sciences and Humanities (CRASSH) of the University of Cambridge for their generous intellectual input and administrative support.

Writing this book would not have been possible without the generous assistance of staff in libraries and archives across the globe. We would like to thank them for their generous help, and especially Arlene Shaner at the New York Academy of Medicine Library, Daniel Demellier at the Archives de l'Institut Pasteur, and Mary Holt at the Matas Library in New Orleans.

We would also like to thank our colleagues at the Visual Representations of the Third Plague Pandemic project, Nick Evans, Branwyn Poleykett, Maurits Meerwijk and Abhijit Sarkar for stimulating discussions on fumigation over the years, and Graham Mooney and Jacob Steere-Williams for discussions on disinfection and medical modernity. Finally, we would like to thank Stavroula K. Koutroumpi for her help with the visual material in this book and Kate Womersley for her impeccable feedback on our ideas.

Earlier forms of the book's chapters have been presented at several conferences, workshops, and seminars. We would like to thank the organizers, delegates, and participants at these events for their feedback on our work in progress. Chapter 6 in this manuscript has been previously published as an article (by Lukas Engelmann) in *Medical History*, and we like to thank Cambridge University Press as well as the journal's editor, Sanjoy Bhattacharya, for the generous permission to reproduce the article in this volume.

Introduction

As the barque *Sophie Goerlitz* approached New Orleans on June 13, 1885, its captain first had to steer the vessel to the inspection station at Port Eads, an isolated area close to the Mississippi Delta, 110 miles from New Orleans. It was the first of three stations through which vessels had to pass before they were deemed safe to disembark in the harbor of the city. At the inspection station, the *Sophie Goerlitz*, like all inbound vessels, was swiftly boarded by a medical officer whose job was to inspect the ship and check its sanitary record. The *Sophie Goerlitz* had left Rio de Janeiro with a load of coffee approximately six weeks earlier. Upon disembarkation, it bore all the signs of excellent health. However, Rio was considered an infected port in 1885, as yellow fever was rampant in the Brazilian city. What was worse, during the six-week journey of the German barque, a seaman had fallen ill with unspecified fever and died en route to New Orleans.

Thus the *Sophie Goerlitz* came under the rules governing maritime sanitation at the great Louisiana harbor: If a ship was known to have departed from an infected port, where an epidemic disease was present and had developed further cases en route, it was subjected to the full extent of sanitary measures to protect the American South. If a vessel, such as in the case of *Sophie Goerlitz*, was known to be "foul"—bearing a case of fever on board—it was directed to the lower quarantine station, located on the Pass A L'Outre. Here, on an unused outlet on the Birdfoot Delta, 103 miles away from the city, the station doctor carried out a thorough inspection of all crewmembers. The sick, if any survived, where removed immediately to appropriate hospitals. The barque was then requested to dock onto the impressive wharf of the second, Upper Mississippi Quarantine Station, seventy miles below New Orleans. The station boasted hospitals dedicated to

yellow fever and smallpox, housing for crewmen and the quarantine officers, and a graveyard, as well as the barracks for comprehensive disinfection
operations. The vessel was brought alongside the wharf and everyone on
board was sent to shore and accommodated in appropriate shelters. Three
procedures were then undertaken to guarantee the sanitary state of the
vessel.

First, all surfaces—with the exception of cargo—were sprinkled with a
solution of bichloride of mercury. Understood to be a universal germicide,
the solution was applied to the bilge, the ballast, the hold, the saloons,
the forecastle, and the decks of the ship. Applied with a simple, heavy,
black-tin rose (similar to an ordinary watering-pot spray), about 2,000 gallons of the solution were used on the *Sophie Goerlitz*. The second—and
most important—step was the complete fumigation of the prepared vessel.
A tugboat was brought alongside the vessel, its pipes extended into the
hatch and, once the fan of the power blower on the tugboat was started,
the *Sophie Goerlitz* was subjected to three hours of sulphuric acid gas exposure. To this end, the tugboat came equipped with a battery of eighteen
furnaces, in which roll sulphur was burned on pans. The furnaces opened
to a common reservoir, in which the power blower pushed the sulphuric
acid gas with huge pressure down the hatchway of the vessel into the bottom of the hold. Within an hour, the brand-new fumigation device on the
tugboat could surcharge the atmosphere of ships with about 180,000 cubic
feet of gas. The third and final step was carried out in parallel to the washing and fumigation of the vessel. All beddings, linen, mattresses, curtains,
fabrics, and personal baggage were brought to a commodious building on
the wharf and treated with moist heat at a temperature of not less than 230
degrees Fahrenheit (110 degrees Celsius). A large chamber had been constructed for this purpose, equipped with racks to loosely hang the fabrics so
as to maximise exposure to moist heat. After completion of the procedures,
the *Sophie Goerlitz* was allowed upstream to unload its coffee cargo onto the
wharf in the city of New Orleans.

The *Sophie Goerlitz* had been one of 210 vessels to approach New Orleans
in 1885 from a port known to be affected by smallpox, cholera, or yellow
fever. All vessels underwent the same treatment, and only after the completion of the three procedures detailed above, and providing that no further
case of fever had developed among the crew and passengers, was a ship
considered as safe to enter the harbor. To the inventor of this new system of

maritime sanitation, Joseph Holt, the process was now in a "nearly perfect state." It was the product of long and tedious "experimental effort" and resulted in the long-aspired and needed modernizing of quarantine, bringing an outdated and questionable practice "into line with other branches of science and art in general progress."[1]

The system of maritime sanitation that Holt envisioned and built in New Orleans marked the beginning of a global drive for the technoscientific modernization of quarantine. Holt's experimental efforts introduced science and engineering to a dated procedure that had relied almost exclusively on detention and observation. Now, enabled through new chemical compounds and uniquely engineered furnaces and power blowers, quarantine was reinvented as a concerted form of intervention that mobilized scientifically grounded fumigation as a comprehensive method of disinfection. At the core of Holt's new system rested a simple gas made from burning sulphur: SO_2. Remarkably, the compound came to be trusted with the total destruction of bacteria, insects, and rodents. In the years following Holt's demonstration of its application, its toxic capacity would foster renewed enthusiasm and harbor the hope for the replacement of the unreliable and costly obstacle to global trade and traffic that was quarantine—a sulphuric utopia.

Histories of Maritime Sanitation

This book tells the history of a sulphur-based fumigation apparatus that was motivated and catalyzed by Holt's system of maritime sanitation. Shortly after its invention in the late 1880s, and its patenting at the end of the nineteenth century, the Clayton machine encapsulated and integrated the complex system set up in Louisiana and saw the widest distribution and application in harbors around the world. Posited against traditional methods, superstition, quackery, and a concerning lack of scientific validity, this new machine mobilized cutting-edge chemistry to reconcile the age-old opposition between commerce and quarantine, and the inseparable linkage of the flow of goods to the rift of epidemics. In this way, its inventors claimed, if adopted globally, the Clayton machine would allow for the first time for a hygienic and secure conduct of international trade and travel.

The machine carried the name of its inventor, Thomas Adam Clayton, a Scottish immigrant born in Banff, Aberdeenshire, in 1852, who settled

in the small town of Opelousas, in the Louisiana Parish of St. Landry, in the
early 1880s as a cotton farmer. Like many farmers in the American South
after the Civil War, Clayton joined the Farmers' Alliance to improve the
economic hardship of rural agriculture.[2] He developed a modest political
career, becoming Secretary for the Alliance in New Orleans, before Govenor Nicholls appointed him in 1891 to the State Board of Health to represent farmers' interests. From there, he ventured into garbage processing
and chemical processing endeavors before he joined forces with the Board's
president from 1894, Samuel Olliphant, in the development of an effective,
industrial apparatus for maritime sanitation.

   Their collaboration led to two registered patents, expanding on the
system that the previous president of the Board of Health, Joseph Holt,
had set out in earlier years. Olliphant and Clayton identified the fumigation apparatus as the most significant contraption within the system of
maritime sanitation and focused on the efficiency and reliability of the
machine. Their product was a mechanical device, attached to a furnace,
which could pressurize and deliver the sulphuric acid gas at constant pressure. The machine could then maintain a continuous circulation between
the furnace and an enclosed compartment, to evenly fill the fumigated
space with gas. As the chemical compound was also shown to have excellent capacities to extinguish fires, Clayton filed his patent on June 7, 1899,
as "Method of and Apparatus for Fumigating and Extinguishing Fires in
closed Compartments." In the same year, Clayton moved to New York to
set up a company whose sole aim was to build and export the very successful and much-requested machine around the world. Yet the development of new procedures, practices, and technologies invented to relieve
epidemic pressure on global trade would exceed the Clayton as a singular apparatus. Already by the late 1890s, rival technologies, employing
different chemicals and methods, competed for the position of the most
efficient technology of maritime fumigation. To compete, the Clayton
would be improved and adapted in response, while also giving rise to new
sulphur-based apparatuses such as the Aparato Marot. Eventually, following
World War I, new fumigation processes based on cyanide came to dominate the field. These, however, continued to carry with them the technical and epistemological legacy of the Clayton machine, whose global
dissemination over the preceding decades rendered it the paradigm of
maritime sanitation.

The Clayton machine inhabits a position in which an astounding variety of historical narratives intersect and intertwine. This is a history of global scientific consolidation in the fight against infectious diseases before the emergence of "global health" as an epistemological and biopolitical framework. The development and distribution of the machine populates a timeframe in between the sanitary conferences of the nineteenth century and the foundation of the Pan-American Health Office in 1902, the Rockefeller's International Health Board in 1916, and later the League of Nations Health Office in 1924.[3] The history of the Clayton machine, its commercial distribution, and its experimental validation presents a world history of science and technology. Its application in quarantine islands and isolation stations across the world places this fumigation technology within a geopolitical context whose reference was not simply the nations and cities that these stations were built to protect, but a global archipelago of trade interests and capitalist as well as imperial competition.[4] Built to advance and support the unhindered flow of goods across the oceans, this apparatus was not only an integral element of a maritime world history. More importantly, it was a technology that enabled a global political economy and drove the success of global trade, thus constituting a forgotten pillar of globalization.

With respect to the inner workings of the apparatus itself, its chemical and mechanical technology was built on the hopes and dreams of laboratory science and sought its validation in the very experimental systems that were being devised at the time to frame bacteria as causes of disease and as agents of epidemics.[5] This then is also a history of the systematic and coordinated utilization of chemical gases as a means of interrupting the natural pathways of diseases, their vectors, and their microbial agents.[6] In this book we move beyond the traditional historiographical focus on DDT and antimosquito campaigns, and examine the Clayton machine and its global republic of epidemiological, medical, and mechanical experts to redraw the timeline of the systematic application of gases and chemical solutions as germicides, insecticides, and pesticides. We ask how the Clayton machine and its technological variants worked to stabilize budding bacteriological science and how its utility in harbors around the world accelerated epidemiological concepts. Finally, we ask how this technology shaped the perception of infectious disease against the backdrop of economic globalization.[7]

The Clayton machine's invention and development crosses in complex ways with the discovery of the causative agents of several diseases and their vectors. First, as we will see in chapter 2, the Clayton's invention and application in yellow fever–ridden New Orleans predated the identification of the mosquito, *Aedes aegypti*, as the pivotal vector of the yellow jack.[8] And yet the overwhelming success of the apparatus in both the American North and South fostered the research of James Carroll and Walter Reed on the etiology of yellow fever in Cuba in 1900, favoring the insect vector over "poisonous effluvia in the air."[9] Second, the Clayton became a significant pillar in global containment efforts toward the third plague pandemic. Erupting in Hong Kong in 1894 and quickly spreading across the globe, plague had been attributed to a pathogen (later known as *Yersinia pestis*) in 1894, but the disease's transmission pathway, maintenance mechanism, and vectors raised many questions. Once again, the machine was tested and entrusted with protection against plague before its paradigmatic target, the rat, had been accepted as a principal vector of the plague.[10]

As much as the Clayton machine connected and associated these two global disease biographies, it occupied a peculiar position within the unresolved etiology of each of these diseases. The apparatus could present its efficiency and capacities against bacteria, insects, and rats, forming the holy trinity of disinfection, insecticide, and deratization (better known as the three Ds: *désinfection, désinsectisation, dératisation*), while each of them were subject to heated controversy about their role in the transmission of yellow fever and plague. So, rather than giving the machine an auxiliary role in the history of the containment of infectious diseases, we need to recognize its substantial position within the larger experimental system that gave these epidemics their modern configuration. Outside of the conventional bacteriological laboratory and beyond the trodden pathways of pioneering doctors, maritime fumigation contributed to the epistemic and biopolitical shaping of these diseases in ways overlooked by current understandings of the history of medicine.[11]

From Infection to Infestation

The examination of fumigation and its apparatuses comes to unsettle prevailing notions in medical and public health historiography in two ways. First, it challenges the notion that by the end of the nineteenth century the bacteriological revolution led to a shift of attention away from objects and toward human bodies. *Sulphuric Utopias* demonstrates this to be a fallacy. For, when bacteriology is considered in tandem with developments in chemical engineering, it becomes evident that while, on the one hand, human bodies enter the scene of public health with unprecedented force at the end of the nineteenth century, on the other hand, object-oriented techniques and technologies of epidemic control also intensify and proliferate. Second, drawing a history of maritime fumigation reveals another misconception: that the advance of bacteriology also led to a gradual shift of attention away from the environment and toward microbes as the true objects of medical intervention. Historians of disease ecology have already shown this to be a perspective that neglects developments of vast importance in the history of epidemiology and public health, including disease ecology.[12] *Sulphuric Utopias* enriches the challenge to this germ-centric approach, showing that fumigation was foremost a practice of both public health and free trade. Within maritime sanitation, it was vital to configure microorganisms beyond a microscopic perspective, and in relation to "macro" spatial and material conditions that could not be more removed from the aseptic environment of the lab: the built environment of cargo ships.

The history told in this book follows the practical consideration that went into the development and distribution of the Clayton machine. However, the practical problems that emerged around the machine's design and application were not simply solutions and fixes, but have themselves also contributed to the conceptual and theoretical development of hygiene, bacteriology, and ecology. This is then a history that begins with the consideration of pathogenic bacteria hiding in the surfaces of walls, ceilings, and floors. Throughout the second half of the nineteenth century, practices of whitewashing had proliferated and—as in the example above—were used to return infected surfaces to a clean state. However, quickly moving beyond the capacities of washing and scrubbing, the Clayton machine fostered an imagination of bacteria hiding in cracks and porous surfaces, as well as in the intermittent gaps and spaces in goods such as tobacco, coffee beans, or fabrics. Its gas was first imagined and then experimentally shown to halt bacterial growth in a great variety of structures, fabrics, foods, and goods. With the consolidation of mosquitoes, fleas, and other insects as vectors of microbial pathogens, this became a history of attributing infection to residues of moisture or bilge water in the holds of vessels, as well as to organisms and insects found in foul air in confined spaces. The Clayton rendered habitats inhospitable to insects, as well as to bacteria. *Sulphuric Utopias* is finally a history of "dead" spaces: the Clayton machine's gas could act in those out-of-reach gaps and structural spaces that enable rats to build borrows or seek refuge and harborage. This is then the paradoxical image derived from the examination of maritime fumigation: while it was initially catalyzed by the bacteriological revolution in the 1870s, it ended up, by the 1920s, being practically indifferent to microbes and concerned solely with their hosts. This is perhaps the most unexpected outcome of this study: that bacteriologically informed maritime sanitation was, historically speaking, a process that gradually shifted its focus from infection to infestation.

Persistent Utopias

Like many technological inventions in history, the success of the Clayton machine is not one of rapid or sudden revolution, but a technoscientific history that is grounded in longstanding traditions, practices, and myths. Traces of the medicinal uses of sulphuric gases go back to Hippocrates and have a vibrant history throughout medieval and early modern periods.[13] Engineering sulphur's germicidal capacities in the late nineteenth century changed the epistemological position of the substance and its use. It did not, however, invent *ex nihilo* its use in the prevention against diseases and epidemics. On the contrary, the Clayton machine claims an almost paradigmatic position in the adaption of an age-old tradition, which was reinvented as a normalized and universally applicable scientific practice.

This is then a history of technological and scientific modernization and globalization. But it would be wrong to address the Clayton machine simply as a technological fix for a series of practical problems made available to sanitary officers at the turn of the century. Beyond its promises of delivering a sound and safe resolution of the enduring quarantine crisis of the nineteenth century, the invention of the apparatus was also driven by a range of imaginaries, desires, visions, and utopias—most pertinently ones related to what in this book we call a "sulphuric utopia."

The notion of "hygienic utopia" has been frequently used by medical historians to describe not just futuristic visions of a disease-free humanity,

as described, for example, in Jules Verne's *Begum's Fortune*, but more pertinently the medical and public health aspirations or programs, ranging from
eugenics to turn-of-the-century mining industry hospitals in the Ruhr.[14] In
relation to the history of epidemics, more specifically, Peter Baldwin has
emphasized the importance of hygienic-utopian visions of modern European states in the development of quarantine as a technology of epidemic
prevention. "Ultimately," he writes, "the goal of prophylactic endeavour
was to sanitise each nation, whether west or east, thus preventing the
spread of disease. But before the happy day of this hygienic utopia had
dawned, much could, in European eyes, be accomplished through the judicious application of quarantinist and, later, neoquarantinist techniques to
the connections between Orient and Occident."[15] In this book we argue
that if indeed this may have been the case in the mid-nineteenth-century
Europe, already by the 1870s the link between hygienic utopia and quarantine was becoming radically unsettled. By the start of the new century
their relation was in fact reversed: hygienic utopia was a state of affairs envisioned as one liberated from quarantine, where the latter's parochial and
time-wasting rules were replaced by the scientific expediency and accuracy
of chemical engineering.

Yet, in promoting this historical revision, we should be wary of an
analytical peril: that of inadvertently carrying with us a largely fossilized
understanding and employment of the notion of "utopia." Medical histories that casually employ the term hygienic or, more broadly, technoscientific utopia often lack a consistent theory of utopia. Or rather, inherent
in their use of the term are contained two evaluations, which, however,
remain nonanalytical. First, in a manner following critiques of utopianism
ranging from Jeremy Bentham to Friedrich Engels, the evaluation that utopias are symptomatic of approaches to social change that are divorced from
the material (or economic) realities on the ground.[16] This is what we may
call the "utopia as magic bullet" approach.[17] Second, in a manner that follows the "antitotalitarian" doctrine of Karl Popper, the New Philosophers,
and their epigones, the assumption that all utopias inevitably lead to, and
thereby contain within them the seeds of, authoritarian dystopias.[18] This
is what we may call the "utopia as the anteroom to the gulag" approach.
Reliance on these approaches, and frequently on their combination, has
led to dismissive and at the same time cautionary historical narratives that
depict utopias as both foolish and dangerous. In this way, a range of less

reductionist and more analytical approaches of utopias and utopianism, which form an important part of twentieth-century social theoretical thinking, are ignored. Most pertinently, this includes the notion of the "utopian imagination" as developed by the philosopher Ernst Bloch.[19]

This book is not a work of political philosophy, but as the result of the collaboration between a historian and an anthropologist it aspires to social theoretical coherence. Hence, while not aimed at reviewing or revising historical or anthropological approaches of utopia per se, we are interested in providing a more clear—and we hope more coherent—use of the term as this relates to hygienic technology. This will in turn help us underline key aspects of the hygienic, technoscientific utopianism shimmering under the surface of the historical fumigation practices examined in this book. And perhaps, this approach of historical utopian visions can also allow a critical interrogation and reflection of contemporary hygienic utopian projects, which, as in the case of the recent Zika crisis in South America, often return to and redeploy spectacles of fumigation.[20]

Hygienic utopias, at least as imagined and performed in the context of maritime fumigation, should be understood within the broader field of what Ernst Bloch's analysis of utopianism has identified as a "concrete utopia"; in other words, a "transformed future" which, however illusive, is practically and materially sought after.[21] Following Ruth Levitas's analysis of Bloch's work, this is a type of utopia that is based on reaching "forward to a real possible future, and involves not merely wishful but will-full thinking."[22] This is, in other words, a vision where "the future is 'not yet' and is a realm of possibility. Utopia reaches toward that future and anticipates it. And in so doing, it helps to affect the future. Human activity plays a central role here in choosing which possible future may become actual: 'the hinge in human history is its producer.'"[23]

The hygienic and technoscientific utopia examined in this book may pale in poetic insignificance by comparison to visions like William Morris's self-proclaimed Scientific Utopia.[24] And yet it shares with them a common principle of "educated hope": the ability to envision an "emergent future" on the basis of a belief that, "the material world is essentially unfinished and in a state of process."[25] In other words, a future that is both radically other to the present and within reach through the exercise of scientifically led human activity. Far from being merely part of technoscientific "dreamscapes of modernity," the fumigation machines and methods examined in this book formed the material, and indeed mechanical, bases of a knowable and realizable future.[26]

Crucially, at least since Plato's account of Atlantis, and in spite of terminologically referring to the imagination or quest for a non-place (*ou topos*), whether abstract or concrete, utopianism has primarily been a spatial fantasy. This is not only in terms of imagining the perfect society as something that exists elsewhere, but also as something that is fundamentally imagined in spatial terms: through its urban planning, architecture, and geomorphic arrangement and transformation.[27] Even after Louis-Sebastian Mercier's *L'An 2440*, when utopia was "temporalized" by being transferred from across the ocean to across time, the spatial aspect of utopianism remained dominant.[28]

In the case of what we call sulphuric utopias, the end of quarantine and the technologies designed to achieve it have entailed a spatial mentality that was entangled with a dreamscape of capitalist hegemony that was particularly triumphant after the defeat of the Paris Commune in 1871 and the inaugural act in the colonization of the global south (Berlin Conference, 1884–1885). Sulphuric utopias undergird the late nineteenth century's economic globalization, envisioning seamless global trade networks built on the imperial maps of Western expansion. The global expansion of capitalist methods of production, the extension of labor markets beyond national boundaries, the rapid growth of global plantation agriculture, and the increasing linkage of markets relied, as Sebastian Conrad has argued, on revolutionary changes in transport.[29] The "amazing decline in international transport costs" was, according to the economic historians O'Rourke and Williamson, the true driver of nineteenth-century globalization. Among other factors, such as the opening of the Suez Canal in 1869, it was in particular the innovations in steamship transport that, a year later, led steam tonnage to exceed sail tonnage in British shipyards. The fall in transport costs ignited a race in the liberalization of trade policy. But, perhaps more importantly, low costs allowed for a transoceanic exchange of "basic" goods, such as wheat, rice, tobacco, and wool. Replacing the previous dominance of special and high-value goods, global trade became structured by the shipment of ever-increasing quantities of food and raw materials from "recent settlements" to the capital-abundant West.[30]

This hegemonic project relied on technoscientific visions of progress and fostered a spatial fantasy that lay at the heart of end-of-the-century

Empire building: the creation of a new spatial order that would guarantee the smooth production and circulation of commodities. This is the political and economic context of the spatial fantasy that, following Douglas Burgess, lay at the heart of maritime capitalism in the age of Empire: the annihilation of distance.[31]

If this is an image often used today, in the age of "jet travel," so as to weave cautionary tales about the spread of emerging pathogens, the trope in itself is in fact the product of the turn of the nineteenth century. The fact is best illustrated by the identical nature of maps used to warn about pandemic danger, where the globe appears to be spanned by a thick web of lines; now airplane flights, then shipping routes.[32] Set at the center of this fantasy was the steamship, which, as Burgess reminds us, Michel Foucault aptly described as a "heterotopia . . . a floating piece of space, a place without a place, that exists by itself, that is closed in on itself and at the same time is given over to the sea and that, from port to port . . . [is] the greatest reserve of the imagination."[33] Seen as a space outside space, but moreover as an indispensible mediator between spaces that mattered (commercial, military, and political hubs) the ship had for centuries been the object of rich symbolic and governmental renderings across the globe. Since at least the Delian League in classical Greece, it had been not simply a carrier of goods and soldiers but also an institutional catalyst of wealth and power. But there was a flip side to this story: for the ship was also the site of mutiny, rebellion, vice, and pestilence. There is no space here to draw the complex genealogy of the ways in which, in Europe at least, ships were constituted into such ambivalent sites. It is, however, important to note that if already in Late Antiquity and then again in the Middle Ages plague was believed to have "arrived" in the supposed seats of civilization (be these Constantinople or Venice) onboard ships, it was only with the rise of concerns about naval hygiene in the British Royal Navy during the eighteenth century that the boat started to be problematized as itself being what today we would call a pathogenic environment.[34] Following Burgess, we can maintain that, by the mid-1800s, "ships were not merely agents of their respective nations, but floating embodiments of empire."[35]

Accordingly, *Sulphuric Utopias* assumes steamships to have materially embodied the space of Empire, insofar as, at least since Pierre Bouguer's 1746 *Traité du navire*, naval architecture reflected not only the practical needs of imperial power (cargo displacement, crew capacity, etc.) but also a set of imperial hierarchies, temporalities, and ideologies.[36] And yet, what Tamson Pietsch has called "the moving space of the steamship" was not always or only that: a moving space.[37] It was also a moored space—a space temporarily attached to the space of quays, docks, and wharfs, or connected to them via smaller vessels while anchored in port. It was moreover often-times a space sidelined and contained in another space, that of the quarantine station, which in turn embodied the transforming and transformative qualities of imperial space in its own way.

Although architectural and, more broadly, spatial histories of quarantine are still disproportional in number to political or social ones, following Alison Bashford, maritime quarantine stations may be said to be spaces of suspension: "quarantine was at once part of the world forged through connections of capital, trade, and empire, and one of the responses perceived to hinder those connections."[38] Krista Maglen has accordingly stressed that, as loci where quarantine was principally applied, ports were "more than just localities for arrival and commerce, they represented sites of tension between notions and constructions of 'exterior' and 'interior' or 'foreign' and 'domestic,' where the dangerous and diseased 'them' encountered the protected sphere of 'we.'"[39] Such sites of tension are notorious for their scattered, incomplete, and often fragmented archives and they present a number of challenges to the historian. Echoing Trouillot, our history of fumigation remains a historical account of scattered actors, and a network of accounts of historical events in which what has happened and is said to have happened remains ambiguous and sometimes inconclusive.[40] Finally, some of the Mississippi quarantine stations, where the Clayton apparatus was devised, have long dissipated with Louisiana's "disappearing coast." Parts of our history of maritime sanitation have thus been lost with that other utopia of controlling the forces of nature that continue to threaten—in the shape of the mighty Mississippi—the very existence of Louisiana.[41]

By the dawn of the bacteriological age, in the second half of the nineteenth century, quarantine had long been employed with the purpose of suspending the circulation of humans and goods. Detention was seen as key in hindering the spread of diseases. According to Erwin Ackerknecht's influential thesis, quarantine merged together commercial and medical debates along the lines of the principal transmissibility of diseases. Contagionists, who held that disease could be transmitted from human to human, argued for strict quarantine throughout the nineteenth century, while anticontagionists opposed the costly restrictions to maritime trade, as they believed disease to emerge *de novo* in the places of its appearance.[42] According to this reading, anticontagionists were hailed as liberal reformers, defending freedom of individual movement as well as the freedom of trade against the shackles of contagionist "despotism." And they were thus often accused as being mere "mouthpieces of a ruthless and economy-minded bourgeoisie."[43] Yet recent scholarship has shown that mid-nineteenth-century quarantine was far more concerned with the question of infection than with that of contagion. Rather than human bodies, the principal source of concern for port authorities was the goods and their capacity of infecting populations at their point of destination. With the concept of an invisible human carrier (such as popularly established later by the figure of "Typhoid Mary") absent until the turn of the century, it was the seemingly inconspicuous state of the goods in the holds of the vessels, as well as the material structure of the vessels themselves, that generated suspicion and necessitated practices and regimes of quarantine, disinfection, and fumigation.[44]

In this sense, quarantine formed a sort of imperial counter-space, where besides suspicious objects that occasionally needed to be destroyed, what was sacrificed was a key component of capitalist production: quantifiable time. Quarantines delayed trade and the movement of both capital and labor power. Although in actual terms, on a local level, such delays may have formed part of complex political and economic circles of profit and power, in terms of both imperial policy and the imaginary goal of the capitalist economy, they were construed as an obstacle. In these terms, the fumigation technologies examined in this book contained a utopian potential that promised to abolish the *stasis* embodied by quarantine as a political, economic, and medical chronotope that, to return to Maglen, "extended both outward into maritime space and inward into port towns and cities."[45] Maritime fumigation was a technology imagined to free up the floating space of steamships from the stone-and-mortar space of quarantine stations and lazarettos. It was, in other words, a technology imagined, engineered, and practiced in and for its ability to halt the circulation of germs without suspending the circulation of capital.

The fact that sulphuric utopias were concrete, and that this concreteness was embodied in an apparatus employed to achieve them, does not mean that these were in some way *limited* visions of a hygienic future. On the contrary, we could say that sulphuric utopias were able to persist across several decades and germinate hopes and aspirations across different nations and empires *because* of the limitations of the material practices employed to achieve them. In other words, and with apologies to Miguel Abensour, if sulphuric utopias were "persistent" this was not because they were imbued with some "stubborn impulse toward freedom and justice."[46] Rather it was because the technologies designed to achieve this state of quarantine-free trade continued to fail, *but not fail too much.*

As we will see in the course of the book, competing maritime fumigation technologies always promised to do what rival or previous models had failed to achieve. At the same time, its proponents always maintained their conviction in the fundamental achievability of the task. Particular goals were renegotiated and displaced, etiological theories were discredited or enriched, all under the confessed-as-realistic goal of complete disinfection. Yet where methods were successful, they were never successful enough; always something came up to disturb technoscientific closure. And in turn, where such methods were unsuccessful, they were never unsuccessful enough; always new scientific or technological breakthroughs came up to reignite technoscientific hope. This combination of *limited success* and *limited failure* created the concrete conditions for furthering the innovation, experimentation, and implementation of technoscientific methods aimed at the utopian *telos* of fumigation.

A Technoscientific Vision

Our history of the Clayton machine is a history of the techniques and technologies of maritime sanitation. As such it enacts doctrines, policies, and principles of maritime trade, national security, and hygienic modernity. As we have already mentioned, the principal fumigation technology under examination in this book enabled the flow of goods and people and hence catalyzed state intervention into trade without posing obstacles detrimental to the livelihood of commerce. But it also reinstated the global territorial orders of the time in terms of hygienic separation: by keeping apart the imagined modern salubrity of the West from the epidemic constituency of the rest of the world. As exposed across this book, the Clayton machine and its reflections and projections can be best understood as an integral element of a larger scheme, a plan or a set of strategic interventions, through which quarantine was supposed to be relegated to the museum of medical and technological history.

Such history of the Clayton can hardly ever be just the history of a technology, or the history of global health, or indeed the history of applied chemistry.[47] Rather, we take the invention, development, experimental negotiation, and successive global distribution of the Clayton machine to be a history in which a political and social economy of maritime sanitation developed against the pressure of epidemic threats and toward the desire for a hygienic, disease-free future. The Clayton machine, and the systems of disinfection in the technoscientific context of which it was invented, applied, and developed, should thus be seen as what Simon Schaffer, David Serlin, and Jennifer Tucker have recently called "sites of complex and socially charged forms of embodied labor and knowledge."[48]

Instruments and technological artifacts like the Clayton machine do not just embody a theory or vision of this hygienic utopia. Instead, as a deeply political entity, of the kind that Simon Schaffer has described, the Clayton developed links with political agendas, mobilized pioneers and defenders of germ theory, enabled proponents of free trade and liberal rights, and emboldened imperial interest as much as it weakened quarantine regimes and their underlying commercial warfare.[49] Furthermore, the Clayton also forged an intimate relationship between the past of globalized approaches to health regulation, as witnessed in the sanitary conferences, and the future of an emerging framework of global health. It hence presents us today with a paradigmatic example of the kind of technological intervention that would eventually come to characterize much of the twentieth-century's regime of global health, built as this is on the foundations of colonial and tropical medicine.[50]

Sulphuric Utopias then presents a story that is as much dedicated to the traditions of a history of technology, as it owes its analytical framework to historical epistemology. We come to think of the fumigation apparatus as an instrument in which epidemic crisis and hygienic utopianism converge. The Clayton machine translated doctrines, scientific principles, and epidemic pressures into concrete material practices, catalyzed by schematic drawings, industrious designs, and experimental systems. At the same time, its application and integration into global trade, quarantine systems, and port authorities consolidated newfound etiologies about cholera, yellow fever, and plague, as their vectors and hosts moved into the focus of pro-phylactic practice. The practical usage of the apparatus encouraged mod-ernized, scientific visions of epidemic disease, while propagating hygienic modernity as state doctrine. As such the Clayton takes up a position of a lever, with which modernization, globalization, maritime sanitation, impe-rial interests, and national identities were forged in intersectional ways.

In recent years, in science and technology studies (STS) as well as in medical anthropology, considerable attention has been put to the exami-nation of what, in their article on the Zimbabwe Bush Pump, Marianne de Laet and Annemarie Mol identified as "fluid technologies."[51] On the one hand, these studies have advanced an understanding of devices as "actors" in broader sociotechnological landscapes. For de Laet and Mol, the colonial water pump under examination is not simply an object with history or histories, but a "fluid object" with agency. Thus sharing a broader under-standing of objects as equipped with agency within network-like relations (as advanced by the dominant STS framework of Actor Network Theory) but replacing the "network" metaphor for a "fluid" one, they problematize machines or devices on the basis of their ability to do, irrespective of their designers'/engineers' intentions or "authorship/ ownership," and instead in relation to their environment of application.[52]

In this book we make no claim that the Clayton or any other machine was an "actor." This is not because we believe that this category should be a monopoly of humans, but because we consider the political philoso-phy underlining the very notion of "action" problematic in the first place. Unfashionable as this may be, we will thus not follow authors like de Laet and Mol (and the broader Latourean paradigm) in reclaiming the agency of the hygienic machines we study.[53] This does not mean that these machines in themselves lacked in fluidity, entanglement, mobility, or any of the other characteristics that within the analytical frameworks of Actor Network Theory seem to qualify nonhumans for agency. These attributes were very much part of fumigation technologies like the Clayton. Although these may be seen as "heroic," in the sense of them being far from the "modest contraptions" imbued with affection (as Peter Redfield has shown) by STS scholars, they were in fact far from "heavy" or cumbersome machines, as they may appear to us today.[54] Instead they were hailed and experienced as highly mobile, trotting the globe on rails and sails, as well as being open to reinvention and adaptation.

Tellingly, the Clayton machine was simultaneously a disinfecting and a fire-extinguishing machine. Moreover, it shared a characteristic identified by de Laet and Mol as key to understanding what they see as fluid technologies: "that whether or not its activities are successful [was] not a binary matter."[55] Rather than examining maritime fumigation technologies from the perspective of if or where they "worked" and if or where they "failed," we should here take Deleuze and Guattari's broader understanding of the machinic aspect of capitalism seriously: as something that works only to the extent that it breaks down.[56] Hence, from our analytical perspective, rather than being passports for some ontological rehabilitation in the narrow confines of twenty-first-century academia, the Clayton's fluidity traits should be seen as an invitation for understanding social, economic, political, and technological processes as they were experienced and performed at the time.

In this study we adopt a historical-ethnographic approach in the sense that the analytical and indeed theoretical conclusions we draw on our subject are drawn from its social reality at the turn of the nineteenth century as this may be reconstituted from the archive. In doing so, we reconstruct the contours of the global experimental system built around the Clayton machine. Our account follows the historical trajectories of French, German, and British bacteriological approaches in tropical medicine and traces their impact into North and South American contexts. The historical geography of the story at hand and its global network is built from maps of references, collaboration, and communication as they emerged from the archives. Without any claim to an exhaustive account of fumigation and maritime sanitation around the globe, this book offers an account of the transnational, transoceanic, and thus global history of developing, testing, and implementing maritime sanitation.

Ultimately, this history forbids its own narration as a story of the modern-day overcoming of epidemic threats, or a saga of the successful installation of enduring maritime stability. Instead, the story told in this book has many disastrous trenches. Chemical warfare would rely heavily on the knowledge established through the experimental improvement of fumigation in the service of maritime sanitation. And, in the end of our story, the Clayton machine was eventually moved out of service and into the archive only when a much more powerful, cyanide-based agent was developed—a chemical that decades later found its atrocious implantation

in the service of the Third Reich: Zyklon B. As is well known, accompanying the particular compound's exterminating properties, a whole range of
terms, metaphors, and practices originally developed as part of the sanitary
processes examined in this book were adapted in the Shoah as means of
concerted dehumanization.

Organization of This Book

In the first chapter of *Sulphuric Utopias*, we draw an outline of the historical conditions that made the Clayton machine thinkable and possible by
the end of the nineteenth century. As a technology, it assembled different
histories and practices, which we sketch out to emphasise the technological
appropriation of older ideas about fumigation, disinfection, and quarantine. Fumigation, the chapter shows, was shaped by a long double history:
on the one hand, as a therapeutic and, on the other hand, as a preventative technology, owing to its palpable association with bad air, noxious
vapors, and miasma. Chapter 1 also examines quarantine, with its roots in
the history of the Black Death, as a cause of concern in the sanitary debates
of the nineteenth century. These revolved around a series of international
sanitary conferences, which brought into dialogue commercial interests,
medical theories, and the role of the circulation of goods and merchandise
in infection. Chapter 1 thus proceeds by dislodging the history of disinfection from its bacteriological underpinnings. It draws a genealogy of fumigation as a practice that was deeply associated with the development of
mechanical contraptions and machines, long before it acquired its modern
significance and bacteriological application.

Chapter 2 explores in depth the historical conditions under which the
Clayton machine was developed in New Orleans, Louisiana. With a focus
on the perceived vulnerability of the city's harbor in the American South
to the onslaught of yellow fever and against the background of a failing
system of detention and suspension of movement of goods and humans,
we describe experiments conducted by members of the Louisiana Board
of Health. Chapter 2 gives an account of how Joseph Holt designed and
realized his vision of a sophisticated system of maritime sanitation in the
1880s. It explores how Holt created an exemplary hygienic barrier against
bacteria, insects, and rodents, while providing a highway to global trade.
The chapter demonstrates how his invention would then be picked up

by the farmer, engineer, and short-time Board of Health member Thomas A. Clayton, who eventually filed his patent for a fumigation and fire-extinguishing machine in 1898, before advancing a global commercial career in the new century.

As an invention on its pathway to global success, by the dawn of the twentieth century, the Clayton machine was confronted by fierce competition in the shape of carbon-based fumigation methods designed in Germany and the Ottoman Empire. Chapter 3 examines how these methods developed around the rat as a probable key vector in the transmission of bubonic plague. The chapter explores the efforts of Pierre Apéry in Istanbul and of Bernhard Nocht in Hamburg to gain control over rats in their visions of maritime sanitation. While the Ottoman Empire persisted in its skeptical position toward any reliable system of fumigation, Germany maintained its own technological invention and continued to protect its harbors and citizens through the application of CO_2, in spite of the risks and problems posed by the invisible and odorless product of Nocht's "Gasgenerator."

Chapter 4 charts out the global success of the Clayton machine and focuses on its integration into what began to take shape as political project of defending Europe against epidemics. The chapter points out how the Clayton was adjusted and grafted to existing disinfection regimes in England and France, and draws out the international politics of standardizing conditions and requirements of disinfection for European trade and commerce. In particular, we examine the neglected role of Ottoman pressure on international trade, via quarantine maximization in the Mediterranean and the Red Sea, and the way this came to charge sulphuric alternatives to rat control with geopolitical implications and consequences.

Chapter 5 seeks to capture the conditions under which fumigation with the Clayton apparatus became a standardized and experimentally stabilized practice. Overcoming differences, aligning chemical, mechanical, and most of all hygienic standards was the key concern of international deliberations at the 1903 Sanitary Conference in Paris. The chapter examines how, despite the failure of reaching a clear consensus, the Clayton continued its pathway to global success and how, in the first decade of the twentieth century, it achieved the character of a gold standard as a technology capable of maintaining the hygienic state of the ports it was applied to.

Chapter 6 examines an applied limitation of Clayton as a fumigation technology and a key alternative that arose in response. Considering the example of Buenos Aires in Argentina as a place in which a technology, designed to protect against the import of infectious disease became adapted and subsequently integrated into the urban fabric, the chapter turns its attention to land-based fumigation. It explores how in unprecedented campaigns, Argentina's hygienists transferred the principle of maritime sanitation into the vision of an urban hygienic utopia. Chapter 6 shows that this relied on the technological development of a key competitor to the Clayton method, the Aparato Marot, or as the Argentineans have called it since 1906, the "Sulfurozador."

Chapter 7 examines in detail the reasons why the Clayton machine's path to global distribution and lasting success was cut short. Since the first decade of the twentieth century, a new compound had fascinated health officers and quarantine officials: cyanide and its gaseous form, hydrocyanic gas. The chapter demonstrated that, as they began to replace sulphur-based systems after World War I, cyanide-based methods represented a symbolic departure from the sulphuric utopia of total hygiene. Proven to be relatively harmless to most bacteria, the increasing prevalence of HCN fumigation marked a shift toward vector control and ecological consideration, leading eventually to the introduction of sprayers and the application of DDT. But on the other hand, its most infamous variant, Zyklon B, would eventually translate the metaphorical content of maritime sanitation, disinfection, and disinfestation into the systematic extermination of millions under the Nazis.

1 Fumigation, Disinfection, and Quarantine

In the course of the thirty years of sulphuric utopias examined in this book, fumigation was reinvented as a practice of disinfection aimed at transforming the routines of quarantine. Starting with the end of the eighteenth century, practices of eliminating bad odors (often by means of smoke) converged with the newly emerging paradigm of germ-free environments so as to create a new barrier to the transmission of diseases across the seas. A tradition of miasmatic and humoral theory, in which sulphur had acquired its own distinctive character, was thus met up with the newly established scientific approach to infection and its isolated agents. This was in turn grafted onto a system of disease prevention, which had been long challenged for its disruptive, costly, and untrustworthy design. Histories of fumigation, disinfection, disinfestation, and quarantine, we argue here, converged at the end of the nineteenth century in the unique configuration of sulphuric utopias. This chapter provides an account of the double history of fumigation as a therapeutic and hygienic technology, the political development of quarantine in the course of the nineteenth century, and a brief overview of disinfection practices and machines preceeding the invention of the Clayton machine.

Therapeutic Fumigation

There is little evidence to support that, with the exception of gynecological ailments, fumigation had been a method systematically employed for therapeutic purposes in Europe before the arrival of syphilis and the application of mercurial vapors for its cure. Mercury and sulphur had long been considered to be the foundational principles of all metals, a theory first evident in Islamic interpretations of the Aristotelian corpus and then adopted by Albertus Magnus in his *De mineralibus*.[1] By the early sixteenth century, however, in a "restatement of humoral theory," Paracelsus set his *tria prima*—mercury, sulphur and salt—as the three principles constituting the human body.[2] Mercury (or quicksilver) was seen as "the principle of activity or liquidity," sulphur as "the principle of heat or organization," and salt as "the principle of mass or solidity."[3] In this fundamentally alchemical perspective, each of the three principia was associated with a type of disease: mercury with putrefactive diseases like syphilis; sulphur with what we would today call metabolic disorders; and salt with disorders like arthritis, which were believed at the time to result from an imbalance of solids and liquids in the human body. Based on the homeopathic principle *"les semblables sont gueris par leur semblables,"* Paracelsian medicine proposed a uniquely chemical cure for disorders and illnesses.[4] In this context, mercury's use as a cure of syphilis witnessed exponential growth, especially in the form of ointments, which antagonized guaiacum as the main therapeutic medium. At the same time, however, mercury also began to be employed as a fumigant for the cure of syphilis.[5] It was not until 1776 and the invention of a fumigation apparatus by Pierre Lalouette that the therapy of syphilis through mercuric vapors reached a technological threshold. Aimed at obviating problems and dangers presented by the practice of mercuric ointments, the internal consumption of mercuric products, and by pan-based fumigation, Lalouette's method was aimed at alleviating the symptoms of the "venereal virus" in the human body.[6] His machine was meant to mark a radical shift from the fumigator propagated by so-called Charbonniere, a "quack" who operated in Paris's Bicêtre and Invalides hospitals by covering the patient's head and inducing him or her to inhale the fumes of what Lalouette reported to be a cinnabar-based substance: "The sulphur combined with mercury, did indeed in burning, set a portion of the mercury at liberty, but the quantity that passed into thro' the mouth and nose, was by no means capable of destroying the virus."[7]

By contrast, Lalouette presented his fumigator as the result of combining scientific experimentation on the production of fumigating powders, diligent preparation of the patient, and the construction of the optimal fumigating apparatus (figure 1.1). The result was the following:

It is an oblong square box, in which the patient is shut up: he sits on an open seat, and his seat is so contrived as to be capable of being raised or lowered as the height of the patient may require. At the bottom of the box, there is a square hole for the admission of the furnace. On a level with the bottom, at one of the sides of the box, there is an opening with a sliding door, through which the powder is to thrown onto the fire. At the top of the box there is another opening which shuts likewise with a sliding door—this is for the passage of the patient's head and neck. That the vapour may be more completely retained within the box, it is right to push the sliding door close to the patients neck, and to put a napkin slightly round it.[8]

Lalouette's fumigation machine was pathbreaking in that it was the first to employ experimental methods, engineering, and design in the production and operation of an apparatus aimed at using fumes or vapors for curing a human ailment. Importantly, like other technoscientific apparatuses at the time, such as the camera obscura, it included human subjects inside the interior of an apparatus, necessitating the assumption of bodily arrangements that may be considered to be part of a broader disciplinary form of biopower emerging in France at the time.[9]

The machine's blueprint and principles were to have a lasting impact in the development of fumigation technologies, especially ones utilizing

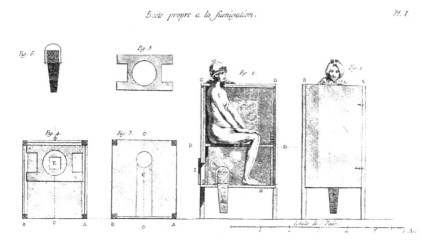

Figure 1.1
Illustration of Lalouette's fumigation machine.
Source: Pierre Lalouette, *Nouvelle méthode de traiter les maladies vénériennes, par la fumigation: avec les procès-verbaux des Guérisons opérées par ce moyen* (Paris: Merigot, 1776). archive.org / Francis A. Countway Library of Medicine.

sulphur. Writing in 1822, some of France's leading medical authorities claimed that, "the application of sulphurous fumigation in the cure of diseases has within a few years excited much attention in Europe."[10] At the center of this medical breakthrough was the invention of an apparatus that allowed sulphur to "be applied to the surface of the body, without the hazard of injuring any of the internal organs."[11] The invention was the work of Dr. J. C. Galès, whose attention was directed to the cure of scabies (*la gale*) while working at the Hôpital Saint-Louis in Paris, an institution exclusively concerned with what at the time were seen as "cutaneous diseases." Already associated by Giacinto Cestoni and Giovanni Cosimo Bonomo with *acarus scabei* in 1685, the disease was laboriously studied by Galès in 1812.[12] The French doctor conducted extensive experiments that showed the disease's causative agent to be the larvae of the *acarus* rather than a "poison" or "virus," while also demonstrating scabies' susceptibility to sulphur. The particular chemical already formed part of therapeutic methods employed against the disease, it being a central ingredient of ointments applied to the human skin before it was exposed to high temperatures.[13] This was a process that, needing repetition over many days, was said to be both very painful and to entail a sulphur-specific side effect: "the skin becomes so saturated with the sulphur, that this foul odor continues to be exhaled for weeks."[14]

Encouraged by the observed impact of sulphur on the acaria larvae, Galès first proceeded with applying sulphuric fumes to patients through the so-called *bassinoire* method:

> The patient was placed naked in bed, every part, excepting the head, being carefully enclosed in the bed clothes, and a pan with burning charcoal, upon which some flowers of sulphur had been thrown, was then introduced into the bed, and the fumes thus brought in contact with the whole surface of the body. By five or six applications conducted in this way, several patients were cured.[15]

Though effective, this process nonetheless risked burning the linen. It also caused serious cough to the patient as a result of the released fumes. "To obviate the inconveniences of the *bassinoire*," Galès proceeded to construct his "fumigatory box" (*boîte fumigatoire*), an apparatus consisting of a furnace and a "box" that was constructed so as not to allow fumes to escape its compartment during operation.[16] The base of the box was formed by two iron plates separated by a few inches. It was on the lower of the two that sulphur was introduced through an opening and was thus carried in an already ignited state inside a small vessel. Inside the main box, and with all openings closed, the fumes of the burning sulphur were carried into the main compartment, where the patient was seated through the iron plates' numerous holes. As can be seen in the diagram provided in Galès's memoirs on his invention, the patient's head was left outside the apparatus, covered with a leather hood, as his or her body was being fumigated (figure 1.2).

Galès explained that he "sought to combine the rules of physics with those of chemistry. By my method, the sulphur is volatilized by heat and thus enters the box at the same time at the caloric; the sulphuric fume covers the entire body uniformly, the face alone being sheltered."[17] As noted by Emerson, in his discussion of Galès's apparatus for *The Philadelphia Journal of the Medical and Physical Sciences*, the vapor entering the box was not

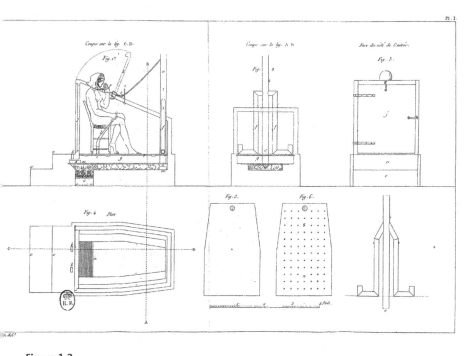

Figure 1.2
Illustration of Galès's fumigatory box.
Source: J. C. Galès, *Mémoires et rapports sur les fumigations sulfureuses appliquées au traitement des affections cutanées et de plusieurs autres maladies* (Paris: L'Impremerie Royale, 1816). gallica.bnf.fr / Bibliothèque nationale de France.

simply sulphur: "During the combustion which ensues, the atmospheric air, which cannot be entirely excluded, supplies a sufficient quantity of oxygen for the formation of sulphurous acid gas"; this was believed "to increase the efficacy of the fumigations" in the human organism.[18]

A key question accompanying this method concerned whether the fumigant operated simply on the surface of the skin and its symptoms or if "the minutely divided sulphur find[s] its way into the circulation."[19] Emerson argued that the latter was the case, as the application of the apparatus to the body, excluding the head, also cured skin disorders on the latter. Indeed sulphuric fumes were said to "give increased energy to the functions of the cellular tissue, stimulate the languid circulation, and modify, in a very remarkable degree, the functions of the lymphatic system."[20]

Galès's invention was submitted to extensive scrutiny by a medical tribunal, including luminaries like Pinel and Dubois, who were so satisfied with the results that they declared it as advantageous over all other existing methods in the cure of scabies.[21] A similarly positive evaluation was secured from a committee set up to examine the method by the Faculty of Medicine in Paris, resulting in its adoption in warships and public hospitals by order of the government. The apparatus was further technically improved by Dr. D'Arcet. This was done by separating the furnace from the main body of the box, and by adding pipes with which to introduce the fumes in the latter, so as to disallow carbonic acid byproducts from affecting the patients under treatment.[22] Soon, an apparatus capable of carrying twelve patients at once was built at the Hôpital Saint-Louis, while the method enjoyed great international acclaim and was widely adopted in hospitals across Europe. Throughout the continent, and as far away as Odessa or the United States, the method was tested on several ailments, including arthritis, herpes, tumors, and paralysis. Physicians like Jean De Carro, the Viennese doctor known for his introduction of vaccination in the Austro-Hungarian Empire, perfected the method and even, in the case of David Luthy in Switzerland, redesigned Galès's machine (figure 1.3).[23]

Fumigation's Double History

The history of the antiscabies fumigator shows us how, already by the beginning of the nineteenth century, technological innovation and experimental science had come together in exploring the uses of sulphur for the

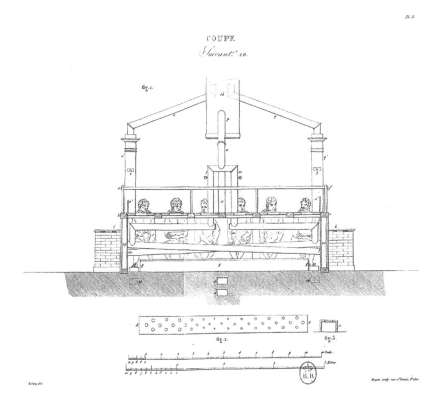

Figure 1.3
Illustration of amended fumigatory box with separate furnace, accommodating six patients.
Source: Conseil général d'administration des hospices, *Descriptions des appareils à fumigations, établis, sur les dessins de M. D'Arcet, à l'hôpital Saint-Louis, en 1814 et successivement dans plusieurs Hôpitaux de Paris, pour le traitement des Maladies de la peau* (Paris: Impr. des Hospices civils, 1818). gallica.bnf.fr / Bibliothèque nationale de France.

benefit of human health. Yet we believe that its history also contains a much more important insight. What is striking is that whereas in the case of scabies, unlike that of syphilis, the causative agents of the disease were known to the inventors, operators, and users of the fumigation technology, the latter was never aimed at *killing* these agents. Instead its imagined impact was believed to be internal—with the fumes said to have a constitutional effect on the patient. This is important insofar as this was the first time that a fumigation technology was used against a disease *in the knowledge of its ability to kill its agent,* and yet this fact seemed to be of little importance to its advocates or users.

This is a case that clearly demonstrates a historical distinction that we need to keep in mind when examining the development and application of fumigation technologies: the distinction between therapeutic and hygienic fumigation. This is notably not a distinction between fumigation practices applied to the human body versus fumigation practices used on nonhuman animals, inanimate objects, or built space. Hygienic fumigation, by which we mean fumigation not aimed at curing but at disinfecting, is used on humans to our days. What defines the distinction between these two modes of fumigation is that the first (therapeutic fumigation) is aimed at individual health, whereas the second (hygienic fumigation) is concerned with communities or, in modern days, "populations," and their health.

If this sounds like a modern distinction, we should point out to a foundational moment in the separation between individual and population/community health. As is well known, this was no other than the Hippocratic differentiation between, on the one hand, individual ailments, and, on the other hand, epidemics. What is less noted, but worth retaining our attention here, is how this distinction relied on the differentiation of a notion that would prove pivotal for the development of certain applications of fumigation in the sphere of human health: miasma. Historians generally agree that miasma first emerged as a magico-religious idea, referring to a form of pollution resulting not from the mere existence of matter out of place (as hasty readers of Mary Douglas may assume) but from ritually inappropriate acts in space and time.[24] Miasma was believed to be a cause of illness and was, paradoxically to our modern ears, experienced as fundamentally contagious: a property or state that was transmitted, among other things, by means of instruments used to purify someone suffering from such a pollution.[25] Yet, at the same time, as Jacques Jouanna notes, miasma and magico-religious medical practice were connected by another phenomenon, not that of individual but of collective illness—*loimos,* or pestilence.[26] As famously portrayed in Sophocles' play *Oedipus Rex,* where the plague of Thebes is caused by the miasma that is embodied by the homonymous tyrant as a result of patricide, the solution to pestilence is also ritual purification, a process portrayed in the *Odyssey* as including fumigation.[27] What Jouanna underlines is that while, as an opponent of such approaches, the Hippocratic corpus ridiculed the individual application of miasmatic etiology (famously in the case of epilepsy, or "the sacred disease"), it actually endorsed and elaborated on it in the case of its collective or "pestilential" application. In Hippocrates's *Breaths*, epidemic fever was seen as caused by miasmata. Used in the plural rather than in the singular, this no longer referred to a ritual pollution but to a natural phenomenon: morbific vapors that enter the human body.

We will not dwell here on the cosmological and indeed political-theological impact of this displacement for Greek society, but rather stay with Jouanna's emphasis on how it affected medical treatment: by shifting practice in the face of epidemics away from purification (and, in turn, from ideas of contagion) and toward the removal of patients from the afflicted location or "environment." What is interesting, from a medical historical perspective, is that, in light of this, in the Hippocratic corpus fumigation was a technology applied not in the case of epidemics but of individual ailments, most famously gynecological ones.[28] It was, in other words, a method used on the human body rather than its environment—a technology aimed at curing the former, and not at halting the progress of epidemics.

Given this extensive historical legacy of fumigation as a practice of treatment, nineteenth-century medical luminaries like Pinel were right in claiming that, "the use of fumigation as a medical agent, is by no means a modern invention, but is almost coeval with the history of medicine." What they overlooked, however, was that, even when considering Europe alone, this use was far from historically uniform.[29]

We can thus already see here what may appear as paradoxical today. On the one hand, fumigation was part of a medical system that gave rise to ideas about the etiological importance of bad air or environmental miasmata. And yet, on the other hand, fumigation was originally radically disassociated from this etiology, and was rather applied to a pathological sphere that was seen as unrelated to the latter. This forces us to reconsider the historicist identification of anti-epidemic fumigation practices in modern times as an etiologically reformed practice or an antimiasmatic technology that was only salvaged from obscurity by an epistemological shift that adapted it to "germ theory." Instead we need to admit that fumigation had always been a far more versatile and multifaceted technology than one defined by or contained in any single etiological framework.

We turn our attention to the hygienic uses of fumigation, but do not for a moment assume this to be a watertight category, unaffected by what we have identified as the sphere of its therapeutic uses. Indeed it may be claimed that if fumigation managed to achieve the hygienic status it did in the decades examined by this book, this was only made possible by it borrowing from and adapting the technoscientific turn already assumed in the therapeutic sphere of its application.

Hygienic Fumigation

The origins of fumigation as a technology aimed at halting or dissipating epidemics are blurry. What is certain is that, in the course and aftermath of the Black Death, fires were used with the purpose of purifying the air from "pestilential stenches," not just with smoke, but also with heat and smell.[30] Such methods, which seem to be based on a hybrid of pre-Hippocratic understandings of miasma as individual pollution and Hippocratic ideas of "bad air" as the cause of epidemics, were endorsed by authorities like King Joao II in his efforts to halt plague in the city of Évora in 1490.[31] In early modern times, as the classical notion of miasma was rediscovered in Europe, plague epidemics were met with the increasing use of public fumigation. Celebrated cases, such as Nathan Hodges and his use of resinous smoke to fumigate plague-stricken houses in 1665 London, largely relied on no more than copper torches, thus demonstrating that by the mid-seventeenth century technological innovation was already evident in antiplague measures. In 1657, during an outbreak of plague in the city of Genoa, the Capuchin friar Maurice devised a simple apparatus in order to purify 430 graves attached to the church, which according to the prevailing medical doctrine at the time were believed to be polluted by the corrupted air emanated by the great number of plague corpses buried within.[32] The apparatus consisted of a large tent built of a wooden frame inside of which was placed a pan containing a flaming substance: a "perfume" made of sulphur, antimony, cinnabar, cumin, pepper, arsenic, and other substances. Two square openings on either side of the tent allowed for a wooden beam to be entered, which would then latch onto a cord attached to the tomb's lid, allowing its operators to lift it. Having filled the tent, the fumes trapped in it were thus forced to enter the sepulcher through its opening and purify it of its pestilential vapors.[33] After one hour of exposure, and after pouring soil on the cadavers, a similar "perfume" was lowered into the tomb with the addition of pulverized sulphur.

If Maurice's method was to all accounts only used once, a few decades later, in the course of the 1721–1722 plague epidemic in southern France, another fumigating machine was invented to deal with the supposedly pestilential vapors of the victims of the disease—one that Sylvain Gagnière tells us was then adopted across Provence.[34] Judging Maurice's method to be both too cumbersome and too dangerous for those operating it, on October 22, 1722, under the directions of the illustrious Avignonese architect Jean-Batiste Franque, a technologically more sophisticated machine was thus built.[35] Embodying both design and engineering in a manner proper to its age and time, this apparatus consisted of a metallic box filled with coals and perfumes; attached to it through an iron pipe were large double bellows, operated manually with the help of a suspended lever.[36] The fumes generated by this machine entered the graves by means of a hole created by a drill. The machine was put in operation in Avignon for at least one month, disinfecting graves until, on November 11, 1722, all antiplague operations ceased.

If these two machines point to a significant output in the early modern development of fumigation technologies against the supposed corrupt air generated by plague corpses, the history of the use of fumigation onboard ships whose crews suffered from disease is more blurry. What is certain is that, with the intensification of medical concern over maritime travel, by the mid-eighteenth century a more systematic study of these methods and their effects becomes evident. This would coincide with and indeed be fueled by a much broader fascination, first scientific and then also popular, about gases and their chemistry. A passion, as Steven Connor has shown, that would engulf Europe in a chemical revolution that relied on the development of an experimental culture and an association of gas chemistry with social reform.[37]

A pioneer of the maritime application of fumigation was James Lind, the Scottish physician best known for his discovery of citrus fruits' effect on scurvy and his work on typhus. In his influential lectures before the Philosophical and Medical Society in Edinburgh in 1761 (published 1763) he relayed how, upon being appointed at the Haslar Naval Hospital near Plymouth three years earlier, he observed the way in which the smoking of the *Revenge* with "the vapour of tar" had led to the abeyance of the infection that had been afflicting its crew after its return from the Mediterranean.[38] Lind further noted that three methods prevailed in purifying ships after the removal of their crew. First, the burning of tobacco, by placing the latter on fires made of old pieces of rope placed across the ship. Second, charcoal and brimstone fire, which Lind noted was effective in "purify[ing] . . . all tainted apartments, ships, cloaths [sic] etc.," but was unable "to destroy some species of vermin, particularly lice"—a fact that led him to conclude "that contagion is not propagated from animalcules."[39] Third, by the addition of arsenic in the charcoal and brimstone fumigants in the following manner:

> After carefully flopping up all the openings, and every small crevice of the ship, (as was also necessary in the preceding processes) a number of iron pots, properly secured, are to be placed in the hold, orlope, gun-deck, etc. Each of these are to contain a layer of charcoal at the bottom, then a layer of brimstone, and so alternately three or four layers of each; upon which the arsenic [sic] is to be sprinkled, and on top of it, some *oakum*, dipped in tar, is to be laid, to serve as a match. The operators, upon setting fire to the oakum, must speedily leave the place, shutting close the hatchway by which they came up.[40]

Lind was confident that the use of fire and smoke (in terms both of heat and fumes) was "the most powerful agents for annihilating infection; and, it may be presumed, even the plague itself."[41] He, however, dismissed the common practice of using bonfires in the streets of cities as a method against plague and other diseases as based on groundless and erroneous principles. He reasoned that such practices were in fact harmful, for they deprived cities of fresh air "fully impregnated with that principle of life" and so much needed "in the inner, foul and pent-up chambers of the tainted sick."[42] If, on the other hand, people were to be removed from such spaces, Lind reasoned, then the benefit of fire and smoke (he did not, it must be noted, use the term fumigation) could not be matched: "In a word, a judicious and proper application of fire and smoke, is the true means appropriated for the destruction and utter extinction of the most malignant sources of Disease."[43]

A year later, in 1762, in the second edition of his essay on the health of seamen in the Royal Navy, Lind provided further discussion of the use of smoke and fire. This included among other things evidence in support of such methods derived from proto-ethnographic narratives. It is here that we come across the first mention of fumigation used for the control of vermin, though these were not associated with any disease at the time. Lind proposed the use of sulphur for killing rats, mice, cockroaches, ants, and weevils. In particular, he advised "that the Fire be at first gentle to draw the Rats towards it, that for they may be stifled in the Hold by the Smoke there, and not at once suffocated by a quick and violent Steam, when dying and afterwards rotting betwixt the Ship's Timbers, they are apt, for a considerable Time afterwards, to occasion a poisonous and noxious Stench."[44]

In 1777, in his groundbreaking book on prisons, the philanthropist John Howard provided a concise summary of Lind's proposed method, derived from their correspondence:

> Charcoal fires should be lighted in the morning, and allowed to remain till evening, and half a pound of brimstone thrown upon each, their smoke in the mean time being closely confined. They may be in iron pots. This fumigation should be repeated every day for a fortnight. Every evening after the fumigation, the ports and hatchways should be opened, and the inside of the ship washed with warm vinegar: and after the last fumigation, before the men return to the ship, the decks should be thoroughly scraped and cleaned.[45]

At the same time, across the Channel, another medical luminary, Louis-Bernard Guyton de Morveau, was the first to experiment with a new method of fumigation on *terra firma*. In 1773 he was confronted with the question of disinfecting the air of Dijon's cathedral. At the time, the cathedral was believed to be polluted with "impure" gases derived from human cadavers kept in it due to the fact that the harshness of the winter season made digging in the church's graveyard impossible.[46] Taking the gases to be the cause of a "contagious fever" in the surrounding area, Guyton sought to counter the pestilential effects of the polluted air by means of muriatic acid (the then-prevalent name for HCl, or hydrochloric acid), in the belief that the latter would counter the "ammoniac" of the putrid decompositions, which were believed to be "the vehicle of the fetid miasmata."[47] Muriatic acid had first been identified as a possible air purifier by Dr. John Johnstone of Worcester in 1758 in a treatise titled *An Historical Dissertation on the Malignant and Epidemical Fever which prevailed at Kidderminster in 1756*. Yet it was Guyton who, basing his work on Antoine Lavoisier's revolutionary understanding of combustion, procured experimental proof of the gas's properties:

I had several times rendered this phenomenon visible, by placing under a very large bell-glass, tilled with common air and immersed in water, two small galli-pots, one of them containing concentrated muriatic acid, or common salt sprin-kled with sulphuric acid: the other ammoniac in a liquid state, or even a solution of the carbonat of ammoniac [*sic*]. White fumes were instantly seen to ascend, fill the capacity of the vessel so as to render it opaque, and then become con-densed so far as to permit the inclosed air [*sic*] to resume its transparency. But it is a fact particularly worthy of attention, that on removing, and replacing it after renewing the air, the fumes will recommence, and the same phenomenon may be produced repeatedly.[48]

Deciding to hold a trial with his fumigating gas on March 6, 1773, Guy-ton used six pounds of salt and two pounds of concentrated sulphuric acid; "The whole was put into a capacious bell-glass inverted and placed on a bath of cold ashes, which were gradually heated by means of a large chafing-dish."[49] So powerful was the gas said to be, that an individual who ventured near the sealed church's main door keyhole was reported as being affected by the escaping fumes. The church was reopened the following day, and four days later services were resumed with the putrefaction having been supposedly extinguished. Content with the results of his trial, Guyton was glad to find an opportunity to repeat it a few months later in the Dijon Jails, where thirty-one inmates had come down with "hospital fever," and then again in 1774, in the course of a cattle epizootic. The latter was aimed at the purification of the affected stables, with the government officially endorsing the use of muriatic fumigation.

It was not long before experiments aimed at the invention of new fumi-gating methods took place in Britain too. In July 1780, James Carmichael Smyth, a physician in Middlesex Hospital, was sent to Winchester's prison where an outbreak of "fever" was raging among Spanish prisoners of war. There he observed that the disease did not follow Thomas Sydenham's rule of epidemic dissipation, but rather increased in force by the day.[50] Unable to observe any petechial spots, as originally reported by the prisoners' cler-gyman, Smyth complained that the "distemper" was particularly deceit-ful, and that even individuals who appeared to be only slightly indisposed quickly perished.[51] Convinced that, in spite of its appearance, the disease was both malignant and dangerous (a judgment seemingly confirmed by himself falling ill with the "fever"), Smyth attributed the outbreak to the general living conditions imposed upon the Spanish prisoners: "in every situation, where a number of people are crowded together, whether in ships, hospitals of prisons, unless the strictest attention be paid to cleanliness and to a free ventilation or circulation of air, a fever sooner or later breaks out amongst them, of a very contagious nature, and attended with very fatal effects."[52]

Smyth entertained the notion that "contagious fevers" were "of two very distinct classes." The first, called "specific contagions," were said to arise not "from any general quality, or process of nature" but to have a "peculiar origin" and thus to "excite diseases of a peculiar kind."[53] Such diseases, like smallpox and measles, were believed by Smyth to be able to afflict an individual only once in his or her lifetime. The second class, called "general contagions" or otherwise "putrid" contagions, was said to result from putrefaction: "That the contagion, or miasma, of the jail and hospital fever," such as the one supposed to be afflicting King's Holds in Winchester, "is of this kind, admits of every species of evidence a matter of fact and of observation can do."[54] Following the long-held opinion that perspiration of bodies held in close quarters was a cause of putrefaction, Smyth generally advocated that the latter might be prevented by the renewal of air. And whereas Smyth, followed John Haygarth in being sceptical about "the risk of propagating the contagion" by means of clothes or furniture lying in an atmosphere where diseases like smallpox prevail, he was confident that this was in fact the case for putrid contagions, like "jail fever" and "hospital fever": "Indeed, wherever a vapor can be distinguished by the smell, we have the demonstration of our senses for what a length of time, not only clothes, but furniture, and even the boards and walls of houses will retain it."[55] Smyth's interest thus lay squarely with a singular question: What was able to destroy the "putrid contagion"? The answer, in his opinion, was "mineral acids in the state of vapour."[56] But while he acknowledged that, in the past, sulphur and muriatic fumes had been successfully used on clothes and furniture, he argued that their danger to animal life made them inappropriate for use in hospitals and prison wards. By contrast, the solution was to be sought in another substance: nitrous acid.

Smyth was confident regarding the efficacy of the particular substance, which was obtained by mixing niter and heated vitriolic acid, on the basis of a private experiment he had conducted (he did not state where or when):

> We put a mouse, confined in a wire trap, under a glass cylindrical jar, capable of holding about 25 pints beer measure, or 881 cubic inches; the jar was inverted upon wet sand, contained in a flat earthen trough or pan; it was then filled with

the fumes of the smoking nitrous acid, introduced by means of a crooked glass tube, until the animal could not be very distinctly perceived. The mouse was kept in this situation for a quarter of an hour, when the jar was removed, and the animal exposed to the open air—it immediately ran about the wire trap, as usual, and had not the appearance of having suffered the slightest inconvenience from its confinement. After a few minutes, the mouse was again put under the glass jar, which was now filled with the vapour of pure nitrous acid, detached from nitre by the vitriolic acid. It remained much about the same time as before, and when the jar was removed, seemed perfectly well.[57]

After repeating the experiment with the use of different minerals on a greenfinch, Smyth finally applied nitrous acid to himself and his friend and collaborator Mr. Hume in a small room (1,040 cubic feet), experiencing no discomfort from the produced fumes.

Although not mentioned in Smyth's own treatise on the Winchester outbreak, Robert Lulman, a Commissioner of the Sick and Wounded Board, recounted the way in which Smyth employed his nitrous acid fumigation in the ward. It had been the first time that Lulman observed the employment of this method: "The vessels in which it was contained were placed on the floor, between the patients beds and in other parts of the wards. . . . The Nitrous Acid was used in a fuming state, while the patients were in their beds, and was kept there night and day."[58]

Following the publication in 1791 of Smyth's treatise on the Winchester events, an official trial of the method was ordered by the Admiralty in November 1795 on His Majesty's hospital ship *The Union*, anchored in Sheerness, which had been recording cases of "petechial fever."[59] Fumigation was applied twice a day on the ship in the following manner, with all patients aboard until February 3, 1796, when the "contagion" was declared extinct: "The articles used in the fumigation are nitre purified and oil of vitriol; which is placed in a pipkin two-thirds full of hot sand, and carried from place to place; a cloud of steam arising from its being stirred, fills the wards."[60] To the great enthusiasm of the Surgeon of *The Union*, during that period only two persons were taken ill. This was attributed to fumigation. Moreover, the process was said to have caused no problems to patient or doctor: "The people bear it exceedingly well, and I frequently stand in the midst of a cloud, arising from fumigation, as thick as fog, without the smallest inconvenience; a circumstance of great consequence, as the sick are all in the wards during the fumigation, and their cloaths [*sic*] &c. are consequently impregnated with the acid vapour."[61]

Used consequently to great acclaim on a number of ships and hospitals (including the Russian ship *Pamet Eustaphia*, the Forton Hospital, and Yarmouth Naval Hospital) Smyth's nitrous fumigation was hailed as both effective and pleasant. This led Paterson of the Forton to describe the method as "of all other remedies extant, the most convenient, the most elegant, the most ingenious, and the most efficacious."[62] In following years, the method was used in a growing number of vessels and hospitals, again with salutary results. It also attracted considerable attention on the international stage, with French and Spanish translations of Smyth's work appearing, alongside endorsements by leading scientists such as Louis Odier.[63]

In 1802 Smyth petitioned the House of Commons for recognition of his discovery. In spite of James Lind's son's objections and an antifumigation testimony by the Scottish naval physician Dr Thomas Trotter, following the consultation of letters and reports from a number of cases where Smyth's method was employed, the Committee of the House of Commons provided its endorsement, awarding Smyth £5,000.[64] The Committee declared that it was convinced "that the efficacy of the Nitrous (or rather Nitric) Fumigation, in preventing the communication of the most virulent contagion, has been fully established."[65] Nitrous fumigation was ascertained to be superior to "other means of preventing infection," such as by use of sulphuric and hydrochloric acids. It was also said to "be used in places the most crowded with sick, without injury to any class of patients":

> Its benefits may be therefore enjoyed, under circumstances wherein it may be extremely difficult or even impossible to maintain the cleanliness and ventilation, which in all cases of infectious diseases are so eminently desirable, and in situations, such as ships of war and transports, where the removal and separation of the sick may be impracticable, and where it may be impossible to cut off all communication between them, and still more, between the persons who attend on them, and others who are in the same vessel.[66]

This was indeed the first official endorsement of a fumigation method by the British government. Yet this success was not bound to England. Encouraged by the reception of Smyth's method and troubled by recent epidemic urgencies in Genoa and in the course of Napoleon's campaign in the Middle East, in 1801 Guyton made a grand return. This took the form of a treatise that by way of thirty-seven experiments reviewed the comparative efficacy of different fumigating agents on "putrid exhalations."[67] Received with great enthusiasm by the French medical community, it elicited the interest of the Ministry of Interior in relation to "contagion" on board of ships, and enjoyed translations in many languages, including English.[68]

Before two years had elapsed, Guyton's muriatic acid method and two fumigating apparatuses, designed by the famous-at-the-time Dumotiez brothers, were not only endorsed by France, but also adopted in Spain in the combat against yellow fever.[69] The machine consisted of a glass bottle contained in a wooden casing, including a wooden cap, which by use of a large screw could be used to loosen the cap and allow gases to escape into the room meant for fumigation.[70] As noted by the historian Elana Serrano, "the fumigation machine embodied two essential features of Lavoiseir's system of chemistry: the theory of acids and the theory of combustion. . . . According to Guyton de Morveau, the fumigating machine supplied a highly oxygenated compound of muriatic acid—oxy-muriatic acid—which destroyed contagious participles in a process akin to combustion."[71]

Serrano has argued that Guyton's idea that "the contagion cannot be born and spread if only by the most culpable negligence" set the stage for fumigation being included within a larger political process, which transformed the relation between the state and its subjects.[72] In Serrano's reading, this entailed a shift from hitherto dominant "military" or "sovereign" measures to more "democratic" ones, which "appealed to citizens' moral responsibility, by asserting their responsibility for their own health and that of their peers."[73] The historical material presented by Serrano do indeed point at an unprecedented public dissemination and state endorsement of fumigation against epidemics. But do they also point out to, or form part of, a shift toward a properly speaking biopolitical paradigm, as Serrano proclaims? Serrano recounts how, in 1804, the Spanish Prime Secretary Manuel Godoy ordered the manufacture of 30,000 such machines for distribution among the general population in southern Spain in order to halt the ongoing yellow fever epidemic. Initially the machines were too expensive and the scheme folded. But after an ingenious redesign that reduced their production cost, they were finally employed a year later "for the 'complete disinfection' of Cartagena."[74] Within the general political atmosphere of reform, anti-absolutism, and the Spanish–French alliance at the time, the apparatus was promoted both in print and by means of public experiments, forming what Antonio García Belmar and Ramón Bertomeu-Sánchez have described as an "imperfect consensus" on the method.[75] This involved its promotion not only as an anti-epidemic technology but also as an enlightened and economically beneficial measure. However, if there is something striking about the emergence of hygienic fumigation at the turn of the nineteenth century, and its dissemination across Europe, this is indeed that, for an age that was increasingly preoccupied with quarantine as a technology of protecting cities and nations from "contagion," the discussion of fumigation seems to be not correlated with the one about quarantine. To thus interpret, as Serrano does, its employment as a process that opposed or overcame the latter would be to superimpose questions and notions from the late nineteenth century to 1800. The employment of Smyth's fumigation in military hospitals and prisons in Britain or Guyton's machine in Spanish households certainly points to a new age when hygienic fumigation would begin to feature among state-endorsed methods of epidemic control. However, as we can see, for example, in Serrano's exposition of the employment of Guyton de Morveau's method in the lazaretto of San Joseph in Cartagena, at the time these were technologies not used to shift the paradigm of what we may call the sovereign power of epidemic control, as embodied by quarantine, but rather to perfect it.[76]

This dynamic between fumigation and quarantine was to change as a result of the rising tensions between proponents and opponents of the latter in the context of consecutive waves of cholera epidemics, beginning with the first cholera pandemic of 1816–1826.

Cholera, Quarantine, and Fumigation

In 1825 the Prussian physician Friedrich Schnurrer presented one of the most eloquent defenses of a geographic view of cholera. His fascinating pamphlet against the existence of any disease independent of the human host—a manifesto against contagion theories—was accompanied by one of the earliest global maps of disease distribution.[77] With this map, and through a vast collection of observations, he demonstrated how various quarantines, put in place throughout the world against the spread of cholera, had failed in achieving their purpose: not a single quarantine had effectively stopped the cholera morbus on its devastating march. To Schnurrer, and to many of his contemporaries, quarantine was deemed to be ineffective for the simple reason that, in their opinion, cholera was not a communicable disease.

However, in the long nineteenth century of disease–etiology disputes, theory and practice often diverged. Only a few years later, the Prussian authorities reacted to the 1830 cholera epidemic with drastic measures aimed at halting the transmission of the disease among districts, buildings, and government departments. A sanitary cordon was put in place around suspicious areas, isolation stations were erected to separate those suffering from cholera, and the authorities set up strict rules for movement across infected areas. Cholera, now thought to be highly transmissible, was supposed to be contained by a drastic regime that combined quarantine and disinfection. Among the measures applied to both the garments and belongings of travelers, as well as goods and merchandise, was systematic fumigation with chlorine. Furthermore, as the Prussian government remained concerned about cholera infiltrating its bureaucracy, every letter, every official *Amtsblatt*, and every file in government possession was supposed to undergo fumigation in order to protect the royal authority.[78]

In the same year, the eminent physician Professor H. Scoutetten of Strasbourg was sent to Berlin to investigate the raging epidemic of cholera. Among his findings, which sought to corroborate an individual disposition rather than a rampant contagion, he presented a new device to treat cholera in individual patients. As fumigation seemed to have enjoyed success in halting the disease in the Prussian countryside, his device proposed a reinvention of the old tradition of therapeutic fumigation for the purposes of epidemic control. The specially prepared bed enabled the treatment of patients with both heat and fumes. A funnel-shaped device at the foot of the bed connected a spirit lamp with four wicks. A tripod and a bowl could be placed over the lamp to vaporize aromatic herbs or sulphur.[79]

For the first half of the nineteenth century, the concurrence of contradictive measures to combat cholera was quite characteristic. Some countries insisted on the effectiveness of strict quarantine, while others focused on local or social conditions. As much as the nature of cholera and its distribution were subject to extensive dispute, so were practices of disinfection and fumigation seen as controversial, with varying prospects of success. After the devastating epidemics of the 1830s, both French and English authorities became increasingly aware of the unsystematic nature of the practices applied. Local quarantine systems came under suspicion of being but masked attempts at obstructing trade. At the same time, intrusive disinfection campaigns led to resistance, as they yielded only very limited effects on the actual spread of cholera. Against this background, and driven by the epidemic pressure of the disease, the first initiatives to arrive at transnational conventions of how to apply disinfection and quarantine were brought forward by the French government. As the second half of the nineteenth century would reveal, the aim was nothing short of sorting, once and for all, the intricate and complicated relationship between quarantine as a system of time-consuming detention, and disinfection as a system of instantaneous chemical protection.

The history of infectious diseases and epidemics has hardly ever been confined to national or regional perspectives. The long history of quarantine, the development of principles of protection through exclusion, isolation, and segregation, shifts the perspective inevitably toward the intimate relationship between local measures of protection and the perception of a dangerous global interconnectedness. Other than hygienic and therapeutic practices of fumigation, quarantine has been an extensive object of study in historical scholarship. As is widely known, the word quarantine stems from the Italian *quaranta giorni*, forty days. It was used in the fourteenth century by the Venetians to describe the length of detention that was deemed appropriate for arriving ships to remain at sea before entering the harbor of Dubrovnik. Despite this original application being driven by the fear of plague and applied to maritime trade, it is crucial to acknowledge that the development of quarantine was neither just a history of infectious diseases nor of maritime sanitation. As scholars have emphasized in recent years, the history of quarantine is an impoverished story, if its deep and integral ties to commerce, economic rationalities, and trade tactics are ignored.[80] Of equal importance is the appreciation of quarantine as a larger system of thought—a system through which a variety of political, economic, and medical institutions, practices of immunization, and structures of regulation have been established on both land and the sea.

The larger environment of the history of quarantine encompasses isolation stations, lazarettos, the long and winding history of the pesthouse, practices of demarcation, spaces of observation and surveillance, and instruments of integration as much as of exclusion. Only recently has Guenter Risse proposed to think of the pesthouse as both a place and a parable in which multiple histories of diseases converge with the historical development of citizenship and public health in the modern USA.[81] Similarly, John Henderson has described the early modern lazaretto in Florence as a place of conversion, where the social and economic structure of Italian societies is laid bare.[82] Howard Markel has portrayed the plight of quarantine in the New York Harbor as a system of exclusion, in which racial stereotypes persisted despite newly available methods of diagnosis and disease identification.[83] And in her account of the "English System," Krista Maglen has shown how the sanitary zones that were crafted around nineteenth-century quarantine stations impacted local and international trade and politics. Throughout history, practices of immunization have yielded to the development of veritable political concepts, whereas figures of immunity align themselves with envisioned communities and their imagined other.[84] Equally, the history of quarantine expands directly from an epistemological history of concepts about contagion to the political story of borders, buildings, and islands to the history of ideas about states, global fabrics, and—crucially—the regulation of trade and commerce.

Mark Harrison and more recently Alison Bashford have emphasized the pivotal significance of the "global archipelago" of quarantine stations as a fundamental condition for the emergence of global trade networks. Harrison wrenched the history of infectious diseases from that of war and the emerging nation state, so as to usher in an almost transhistorical conflict of interest between freedom of trade and the control of disease.[85] For a long time after plague established this paradigmatic conundrum, the world was clustered along the lines of those favoring rigid quarantine for the cost of commerce, and those—like England—who protected their liberal regime as a promercantile, antiquarantine conviction. Harrison thus traces the spread of disease "along the arteries of commerce" so as to foreground the establishment of quarantine systems as corridors of power, in which the liberation of trade was negotiated against the threat of disease.[86] Bashford, on the other hand, sees the global network of quarantine stations coming into new historical light against the development of a perspective in transoceanic history. An icon of world history, the quarantine station thus assumes its significance precisely in its locality, separated from the place of protection but organized as a gateway for global intersections and connections. As a "portal" to world history, the vast network of stations, lazarettos, and ships built to regulate the relationship between trade and disease at the threshold of local and international harbors invites careful interrogations and comparison of the *longue durée* of the geography of diseases, politics, and power.

For both Harrison and Bashford, the history of quarantine cannot be disconnected from the emergence of globalization and of economic markets that transgress the nation-state, nor from the development of an early version of international, or rather, imperial hygiene.[87] Reinvigorating the contours and borderlines of nation states, quarantine stations became thresholds of the political and geographical authority of empires in the late eighteenth century. At the same time, Harrison argues, we see the development of a concept of balance to structure the relation of trade and contagion. Trade interests, along with early humanitarian sentiments, began to define the contours of early global agreements—or rather alignments— about the importance of disease control in the shadow of economic interests. And here, as Ackerknecht has made clear, a new ideological frontier emerged, in which the rationality of quarantine, the particularities of its implementation, and the coherence of its systematic application became subjects of heated controversy.[88] This approach would establish two new guiding principles that structured the majority of nineteenth-century quarantine stations. First, with the 1851 Paris sanitary conference, an international system of corroboration and collaboration was set up in order to align international policy and practice. And second, these new channels of early international considerations of questions of health, disease, and trade would become pipelines for a new kind of political authority, which aimed to make quarantine scientific while asserting imperial economic interests.[89]

The history of nineteenth-century international sanitary conferences revolves largely around the changing particulars of cholera. But the political pressure from which the impressive and unprecedented initiative of global cooperation developed was erected on a shared frustration toward existing practices of quarantine. The British diplomatic delegate to the 1851 sanitary conference, Anthony Perrier, summarized the concerns of those participants, who had no faith in the possibility of cholera's importation. He condemned the quarantine established in some states against the disease, as this continued a "routine path of practices that are outmoded, useless, ruinous to commerce, and harmful to public health."[90] If the majority of the delegates supported a ruling in favor of quarantines against cholera, the result was never ratified by any of the participating states, and was never turned into practice. If anything, the first as well as the second sanitary conference, in Paris (1858), demonstrated the deep epistemological divide on the etiology of cholera. An agreement on standardized quarantine systems was thus out of reach.

By 1874, Russia's frustration over arbitrary quarantine detentions by the Ottoman Empire led to a fourth conference, this time held in Vienna: "practically the whole shipping trade of Russia passing through the Bosporus has, since 1866, been at the command of the quarantine agents of the Porte and their fancies."[91] Although numerous meetings were dedicated to the formalization of international conditions for maritime sanitation, the differences between those preferring quarantine and those favoring medical inspection seemed unresolvable. As the eminent German medical geographer August Hirsch proposed, it was agreed by all to consider all measures viable to the protection against cholera.[92] States were thus free to choose whichever procedures they preferred. In the meantime, Robert Koch's discovery of the causative bacillus of cholera boosted the faction favoring the communicability of the disease. Still, an international agreement was far from reached. In 1892, the seventh international sanitary conference, in Venice, came close to establishing a consensus on the mode of destruction best suited for cholera and, for the first time, delegates were moved to consider the success of steam sterilizers on ships. Still, disagreement prevented a unified proposal for methods of destroying the "pathogenic germ," as sterilization of merchandise and the hulls of vessels did not tackle the issue of the human carrier. Where the conference succeeded was in proposing a classification of ships as "clean," "suspect," or "infected," a classification that, although it remained in place for the decades to come, caused extensive debates over the definition of its categories.[93]

Disinfection: A New Frontier

It would take until the tenth international sanitary conference, which took place in 1897 once again in Venice, for discussion between the world's great powers to consider diseases other than cholera, with the conference being exclusively concerned with plague. This was also the first time international delegates would take into account the necessity of a unified strategy of maritime sanitation across national and etiological divides. In Venice, plague was discussed with surprising unanimity. Although outbreaks of the disease in North Africa, the Middle East, and Russia had not ceased throughout the nineteenth century, it was the 1894 outbreak of bubonic plague in Hong

Kong that brought the disease back on the international stage. As by 1896 plague proceeded to spread across British India, the fear of a global pandemic made it the central subject of the conference.[94] Delegates accepted that the disease was caused by a bacterium, as Alexandre Yersin had shown in 1894, and that outbreaks of human plague were often preceded by epizootics in rats, but not that rats were the actual source of the disease.[95] Still, the role of insects remained ambiguous, and although the American delegates pointed toward the significance of yellow fever, the disease was hardly discussed. The conference succeeded in establishing an internationally binding convention of dealing with plague, adopted by the eighteen participating states and their corresponding port authorities. This detailed the measures deemed necessary to halt transmission over both sea and land. The quarantine for ships suspect or infected with plague was set to ten days; rigorous measures of disinfection were made obligatory for suspect and infected vessels; regular inspections were encouraged, especially for ships dedicated to pilgrimage, and the sanitary station at the Suez Canal was improved and modernized.[96]

The French delegate Adrien Proust considered the regulation of disinfection to be the most important achievement of the conference. In his detailed report, the practice of disinfection was now globally defined as a suite of escalating measures: whereas fire should be used for objects of no value, steaming was decreed as best to clean fabrics, washing with lime-solutions for surfaces and walls of inhabited places, and finally, aeration of spaces with liquefied sulphurous gas was proposed. "It has long been thought, and it is still hoped," Proust wrote, stressing the necessity of fumigatory equipment, "that the best method of letting an antiseptic penetrate everything is the controlled release of a microbicidal gas."[97]

It thus took until the Venice conference in 1897 to arrive at the foundations of an international agreement about systematic measures of disinfection to be trusted with the task of aiding and, in certain circumstances, replacing quarantine stations. It is noteworthy that most of the agreements reached in Venice were exclusively bound to the question on inanimate objects and their relationship to the transportation of disease. In the case of plague, this extended to animal vectors and the possibility of the rat being a carrier of the disease, but by and large, human bodies were taken out of the equation. These bodies remained subjugated to detention times, while goods and merchandise were increasingly considered to be safer if treated with extensive disinfection. Accordingly, the measures agreed upon at the 1897 conference were solely directed at objects and their surfaces and textures, as well as agricultural products.

This is not in itself surprising as recent scholarship has shown that throughout the nineteenth century debates around and practices of disinfection were singularly focused on inanimate objects, fabrics, and goods. Since the first sanitary conference in 1851, objects of daily life, and especially those involved in maritime trade, were categorized, scrutinized, and tested for their susceptibility to infection of any kind. To this end, infection was not clearly defined and was subject to extensive controversy. And yet, as David Barnes has argued, it was precisely this mutable notion of infection that allowed quarantine doctrines and practices to persist across the globe within a timeframe where epistemological and political rejections of notions of contagion were peaking: "What worried medical men and laypeople alike was not contagion but infection, in its pre-germ-theory sense: an invisible, often (but not always) malodorous contamination, potentially transportable over long distances in goods, vessels, or bodies, that was capable of causing disease unless neutralized or dissipated."[98] Long before bacteriological paradigms were applied to define the role of fabrics, organic matter, and inanimate objects in the retention or transmission of agents of disease, these had become pivotal targets of disinfection as part of what Barnes calls an "interpermeable world."[99] Regardless of the failure of sanitary conferences to arrive at an agreement about the usefulness of disinfection until 1897, Barnes has shown for the case of the lazaretto and quarantine station of Philadelphia in the United States that disinfection practices and the application of various chemical substances had long been considered to be effective against "infection," which was often, but not exclusively, understood to be carried (or indeed self-generated) in fabrics, coffee, tobacco, or any other goods that crossed the oceans.[100]

The historian Owsei Temkin has considered the nineteenth century as having brought a secularization of the age-old concept of infection. In his account, the term had been redirected to acquire a scientific content and to develop a "new moral force."[101] Replacing older notions of staining and impurity, according to Temkin, the new medical concept of infection emerged steadily in the course of the nineteenth century through various concepts, proposals, and theories that eventually led to the work of Pasteur, Lister, and Koch. Barnes's portrait of disinfection in the United States suggests there was indeed a "tacit consensus about infection"; a rarely speci-fied but widely applied category that was considered both a process as well as a quality that was bound to certain local conditions.[102] Much of its verac-ity, Barnes proposes, was earned through it being applicable to polemics in favor and against the idea of the contagious transmission of diseases. But unlike in the positivist reading of Temkin, Barnes sees the concept of infec-tion as remaining largely undefined. He furthermore suggests that much of the particular notion's impact on the early landscape of global health was owed to its status as an uncertain object at the center of practices and strategies of containment: "Diffuse and invisible but deadly, infection in the pre-bacteriological era can ultimately be defined only as that which was neutralized or removed by whitewash, chemical solutions, fumigation, steam, ventilation, and flushing with fresh water."[103]

The history of disinfection, which developed over the nineteenth cen-tury, both in tandem with and in disjuncture to quarantine, is often curbed of its rich implications and entanglements, when looked at only through the lens of its bacteriological reinvention in the 1870s. Historical scholar-ship has overwhelmingly focused on the second half of the century and the impact of bacteriology on the reinvention of the field.[104] Yet, as indicated above, already in the first half of the nineteenth century, we see a large number of experimental procedures put in place to evaluate the effects of chemical and organic substances for the purpose of disease prevention. Var-ious indications for the widespread disinfection and fumigation of mail in the early nineteenth-century containment of yellow fever outbreaks in the northern United States demonstrate another field, in which disinfection practice was thought to have substantial impact.[105] In contrast to the above-discussed development of hygienic fumigation practices up until 1800, the question that arises here is at what point the application of disinfectants became a practice in which a chemical agent was understood to have a direct impact on the properties of an infectious agent.

Andrew Mendelsohn argues that much of the early bacteriology in Paris should be seen as focused on the chemical manipulation of the virulence of pathogenic agents.[106] Pasteur, in his own original research on wine, had investigated the possible role of microorganisms in the process of fermen-tation. To halt the spoilage of wine, in 1861 Pasteur then famously pro-ceeded to suggest three different methods of eliminating the responsible microorganism: filtration, exposure to heat, and application of a chemical substance. For the latter, he experimented and ultimately relied on a substance that had been discovered in 1834 by Friedlieb Ferdinand Runge, originally called Phenol and later established as carbolic acid. The substance, which was derived from coal tar, had been used to treat railway ties, as it appeared to prevent rotting. It had also found various uses in sewage and garbage processing, as it seemed to prevent the production of supposedly noxious gases and stinks. Influenced by Pasteur's research on fermentation in wine, it was Joseph Lister who famously began to apply the chemical as a disinfectant on the human body in surgery. He began to wash instruments, soak pads, and clean hands of the surgeons with a 5 percent solution of carbonic gas and observed a drastic reduction in the incidence of gangrene, the most common cause of death after surgery in the nineteenth century. In a series of six articles in *The Lancet*, Lister published his results in 1867 and proposed the paradigm of antiseptic surgery, which—after some controversial back and forth—became rapidly standardized across European and North American operation theaters.[107]

The American physician Robert Bartholow summarized disinfection practices developed over the mid-nineteenth century in his 1867 treatise on "the principles and practice of disinfection."[108] Bartholow argued that the ravages of cattle plague and the disastrous experience of cholera since the 1830s in North America and Europe sparked a renewed interest in disinfectants as weapons against the "mysterious animal poisons" that seemed to be driving the epidemics. The "strictly modern science" of chemistry had supplied society with numerous new agents that could be employed for systematic disinfection.[109] These agents were supposed to show effect against two entirely different categories of substances against which they were employed. First, the products of putrefaction, which created a powerful morbid agent, were considered. Second, Bartholow wrote of a "peculiar organic matter: to which we apply the terms virus, *materies morbi*, morbific matter, diseased germinal matter."[110] When it comes to disinfecting agents, he saw the field as divided between those of chemical nature, employed to destroy a noxious compound (heat, ozone, chlorine, bromine, iodine, nitrous acid, and sulphurous acid), agents such as antiseptics and colytics, which were useful in arresting the chemical change of morbid matters, and agents that were used to physically restrain the noxious substance. Sulphurous acid was to Bartholow in 1867 a common deoxidizer, which had the capacity to abstract oxygen to pass on to sulphuric acid. As such it ranked

high and was understood to be both a brilliant destroyer of infectious matters, as well as a high-functioning antiseptic. As had been long observed in
the age-old tradition of winemaking, sulphurous acid "prevents change,
decomposition, or fermentation, by abstracting oxygen and by destroying
those minute living organisms necessary to this process."[111]

In the decades following Bartholow's publication, and in parallel to the
success of bacteriology, disinfection became a blossoming field of experimental study and theoretical exploration. But again, reducing the success
of disinfection practices to the development of the bacteriological laboratory would miss a crucial point: the continued application of disinfectants
was not inherently bound to a clear definition of the microbe it was supposed to act upon. Rather disinfection was itself an experimental practice
through which the shape and nature of infectants was supposed to be illu-
minated. Rebecca Whyte has argued that, for this reason, disinfection did
become a popular public health practice in the second half of the nineteenth century in the United Kingdom in, as well as outside of, the laboratory. Day-to-day disinfection did mostly take place outside of laboratories.
And although the bacteriological paradigm achieved a model character in
explaining the efficacy of disinfecting agents, "there were myriad debates
about how to create a mutually understandable scientific standard for testing" the compounds on the "epidemic streets."[112] Where the laboratory
and "germ theory" succeeded in redefining and reframing disinfection as
a "germicide," a range of methods remained in place despite the inability
of the laboratory to prove their efficacy against germs, microbes, or any
infectious matter.[113] Whyte explains this historical disjuncture through the
differences between laboratory practice and practical application. Missing
a theory that would sufficiently explain how chemical substances effected
bacteria, rejecting the comparability of the laboratory and the interior of
vessels led to a research field that was in its entirety carried by an experimental system that operated vastly without a comprehensive theoretical
understanding of the supposed efficacy of gases, solutions, and substances
in the field.

In the last two decades of the nineteenth century, this development
opened the floodgates to a vast amount of experimental setups, which were
put in place on both sides of the Atlantic to establish standardized hierarchies of disinfecting substances, as well as to arrive at a consensual theoretical understanding of how modern methods of disinfection were supposed

to overcome all shortcomings of traditional practices of quarantine detention and medical observation. What spearheaded this process was the rapid
development of a range of disinfection apparatuses.

Disinfection Apparatuses

The history of the development of mechanical contraptions for disinfection spanned the nineteenth century. Back in the 1830s William Henry
had conducted the first scientific experiments with disinfection, as a result
of rising anxiety in his hometown, Manchester, about the importation of
plague from Egypt inside bales of raw cotton and subsequent fears of the
arrival of cholera in England's key merchant hub.[114] In his letter to the
Philosophical Magazine (October 14, 1831) Henry drew authority from Alexander Russell's *Natural History of Aleppo* (1756, and revised edition by his
brother Patrick, 1794), where summer heat was considered as a palliative of
plague and which had an impact on Henry's ideas of disinfection.[115] Henry
proceeded to test the effect of heat on smallpox vaccine lymph, which
was neutralized after being exposed to 140 degrees Fahrenheit, while also
showing that "items of clothing made of cotton, silk, wool, and fur were
unharmed after exposure to a temperature of 180F for three hours."[116] He
concluded that a temperature of 212 Fahrenheit was the minimum "capable of destroying the *contagion* of fomites."[117] Graham Mooney has argued
that Henry's experiments "provided no more than analogous inference
for the action of heat," so that, by "directing his lymph-contagion toward
the impending cholera threat, Henry recommended the construction of
dry-heat chambers at seaports."[118] Henry envisioned already in the 1830s
that his apparatus (an enlarged metal boiler, which he soon modified, "for
cholera-hospitals, lazarettos and stations where large quantities of articles
are intended to be disinfected") would, if installed in every port, eventually overcome the unreliable and widely mistrusted practice of quarantine
detention.[119]

   Henry defended the necessity of his method on account of the need to
"substitute" quarantine, and in the knowledge that substances like chlorine
were inapplicable as disinfectants to a wide range of products, fabrics, and
goods as it discolored and damaged them.[120] By the 1870s such methods
would be widely employed in the construction of disinfection apparatuses,
which, as Graham Mooney has shown, enjoyed great popularity. Following

Mooney, in 1870 George Fraser's disinfector, "was designed as a brick oven, heated from underneath by a coke fire. The iron carriage used to transport the clothes was driven directly into the disinfecting chamber, meaning that employees did not have to handle infected clothing and the carriage itself was disinfected during the disinfecting process."[121] What was crucial to these machines was their ability to be "self-regulating," a task confounded, Mooney tells us, by the fluctuating temperatures in the disinfecting chambers when in operation. Though dry-heat machines that approached this ideal were in fact achieved (see for example William Ransom's disinfection apparatus, 1870), the inability of heat to penetrate denser or larger items led to a move toward steam as the preferred medium of disinfection.

Mooney's research on the history of such apparatuses confirms the opinion that this shift was catalyzed by the endorsement of steam or "moist heat" by Robert Koch in 1881.[122] Steam was not only more evenly distributed in the disinfecting chamber than dry heat, it could also penetrate objects thoroughly at lower temperatures. Moreover, a key element to the apparatuses that ensued was their portability.[123] Machines like Washington Lyon's steam disinfector, patented in England in 1880, "had wheels and could be carted around by just one horse."[124] Mooney's study of the history of disinfection in England has stressed how confidence in heated steam was fostered by experiments conducted by Franklin Parsons for the Local Government Board in 1884: "Parson's three-way experimental matrix of bacteria, disinfection method, and materials established a sort of sliding scale for disinfecting efficiency. Using both laboratory methods and operational disinfecting machines in situ, he assessed the impact of dry heat, boiling water, current stream, and steam under pressure on anthrax, swine fever . . . and tuberculosis."[125] Parson's experiments showed steam under pressure to be the most efficient method of disinfection, leading to the adoption of Lyon's machine, and at the same time to the invention and manufacture of competing steam-based disinfecting apparatuses.[126]

In England, the principles of "easy use, portability and effective means of destruction" were linked to experimental systems that came to merge and entangle laboratory and domestic spaces.[127] At the same time, the development of regimes of maritime fumigation required the inclusion of vessels into experimental systems that form the subject of this book. Fumigation, with its long and versatile history of cross-fertilizing humoral and hygienic ways of thinking about the body and its environment, merged in the last three decades of the nineteenth century with the increasingly biopolitical landscape of disinfection. The vapor, known for centuries to be therapeutic in supporting bodily balance and to promise a fortified, hygienic atmosphere against ailments, was slowly transformed into a chemical substance with dedicated effects on organic forms of life. As fumigation became a method of disinfection, it advanced into a powerful agent that enabled and encouraged the possible replacement of unreliable and controversial quarantine routines by a thorough system of maritime sanitation, which promised for the first time enduring safety to fuel a sulphuric utopia of trade and travel free from disease transmission and time-consuming detention. Never totally disentangled, but retaining a material and epistemological exchange allowing for their relative autonomy, by the end of the nineteenth century, domestic disinfection and maritime fumigation regimes formed part of a broader biopolitical, political-economic, and imperial world system that assumed the chemical battle against infectious diseases as a key goal of technoscientific, economic, and social transformation.

2   The Birth of the Clayton in New Orleans

Thomas Adam Clayton settled in Opelousas, Louisiana, some time in the early 1880s. The son of a doctor's family from the Scottish North had immigrated in 1878 into the post-bellum United States to become a cotton farmer. The St. Landry parish, in which Clayton acquired farmland, was dominated by vast prairie land, with few scattered farms, all of which struggled in the uncertain economic period after the war. Clayton soon joined the surge of the Southern populist movement, organizing and representing farmers' interest in reducing farming's economic uncertainty. Over the 1880s he embarked on a modest political role in Louisiana as secretary of the Farmers' Union. In 1890, the Governor of Louisiana entrusted him with a lay position in the state's Board of Health, where Clayton went on to join the ongoing campaigns of safeguarding New Orleans against yellow fever.

Violent waves of yellow fever had rolled over New Orleans in regular intervals, and mortality rates in the first half of the nineteenth century spiraled up to 8,000 per year. In 1853, yellow fever accounted for half of the city's annual mortality.[1] As reports of dead bodies lining the streets, benches, and public places mounted, tradesmen began to search for acquired and innate immunity, often underlined by racial divisions.[2] Meanwhile, the enduring debate between contagionists and anticontagionists went on to shape nineteenth-century reflection on yellow fever in the American Tropical South.[3]

A praising report on the history and present situation of the Louisiana State Boards of Health, published in 1904, following the establishment of the yellow fever etiology and its vector, *Aedes aegypti*, presents an intriguing outline of the state's attempts at the time to bring this enduring epidemic crisis under control. The report voices the firm opinion that the time prior to 1850 would best be characterized as a period of long apathy. Patchworks of different containment strategies, aligned with various theories and concepts about the nature of yellow fever, were applied with more or less success. The earliest traces of applying quarantine practices to halt spatial transmission of the fever were reportedly led by William C. Clairborne, a local physician who already at the dawn of the eighteenth century installed a mechanism to "subject the shipping entering the Mississippi river to those quarantine regulations which in other ports had proven salutary."[4]

The city invoked preliminary quarantine laws in 1816 and 1817. These were scrapped in 1819 so as to delegate issues of quarantine to the state Governor, who in turn installed the first quarantine station, indeed the first in the United States, in 1821. The station was located on the Mississippi River, sixteen miles south of the city, and was mostly used for checking ships and travelers for visible signs of illness and fevers. "It is probable," the report stated, "that some form of 'purification' was also practiced."[5] Initially celebrated as a triumph, the station soon lost its appeal. Yellow fever continued to plague the city and its trade, and the intrusive and costly practice of quarantine appeared as an unnecessary burden to the city's commerce.

As early nineteenth-century attempts to keep the fevers out by presenting physical barriers to its "poisonous" agents appeared to have largely failed, proponents of anticontagionism gained traction. They vocally argued that the disease was of local origin, and that it spontaneously emanated out of the moist grounds of the city and its surrounding swamps. They moreover believed that its human-to-human transmission was as impossible as was its transport in merchandise across the sea.[6] This argument was aptly supported by the costly toll quarantine took on the commerce of New Orleans.

"Yellow fever," Margaret Humphreys points out, "was a profound burden upon southern commerce."[7] This included losses from the direct impact of outbreaks as much as costs incurred from ships, tradesmen, and customers being held up in lengthy quarantine detention. Furthermore, Humphreys reports irregular quarantine checks on the railways, where arbitrary decisions were made regarding trains being disallowed from stopping in New Orleans. As the question of the disease's transportability was deeply associated with the question of commerce, opponents of contagious understandings of yellow fever were frequently, but not exclusively, seen as mouthpieces of commercial interests.[8] Yet, according to Humphreys, a noncontagious view of the disease was also shared by most southern physicians up until the 1850s. The disease's predominant appearance in towns and cities was explained by means of various theories, ranging from "overcrowdedness," to the number of living or dead animals, or the presence of unremoved piles of human excrement, and the constant upturning of soil during construction works, which exposed putrefying matter.[9] These largely miasmatic understandings of the fevers were challenged with an increasing awareness for the successive spatial distribution of outbreaks, when, for example, cases in smaller surrounding towns appeared regularly in short intervals after a larger outbreak in New Orleans. Still the idea of transmissibility in vogue during the 1840s was based on a concept of infection that kept its distance from the traditional understanding of contagion. If yellow fever could be carried by persons, it was believed to be spread not directly from them, but through their clothes or belongings, as it was believed that merchandise, material goods, and trapped air could be carrying the disease.[10]

Throughout the period and up to the middle of the century, quarantine systems were mostly retired, and, for a short period of time, completely halted. Whether or not the complete lack of preventive measures did indeed lead to the largest outbreaks of yellow fever in New Orleans history is subject to controversy.[11] In 1853 and 1854 over 10,000 people succumbed to the epidemic. Initially, medical professionals, under pressure from tradesmen, kept the public unaware of the spiralling mortality numbers. It took until July 1853, with already over 1,000 deaths from the fever, for the need to announce the ongoing epidemic and for the subsequent installation of a restrictive quarantine to finally outweigh commercial and political interests.[12]

The mid-nineteenth-century outbreak was widely regarded as a turning point for the southern port and its strategies of containing the fevers. In the aftermath of the devastating epidemics, the state of Louisiana went on to install the foundations of an ambitious system of disease surveillance and epidemic control in the South. The Board of Health in New Orleans—the first institution of its kind in the United States—had already been in place since 1804.[13] As a reaction to the large outbreaks of 1853 and 1854, a new State Board of Health was founded in 1855, formed of nine competent citizens, who needn't all be physicians but were required to refute theories of the local origin and spontaneous emergence of yellow fever. Six of them were named by the state governor, and an additional three by the New Orleans City Council.[14] The Board would govern the Southern response to yellow fever as an imported and foreign threat until the end of the century. Its influence was fostered by three consequent years of appalling yellow fever mortality. These had shifted opinion in large parts of the profession and the general public, fostering the conviction of a foreign origin and thus of a principal transmissibility of the disease. The old and defunct quarantine station was considered a faulty system due to a lack of control and rigorous protocols, raising additional concerns due to its proximity to settlements along the river. A new quarantine station was installed not less than 70 miles below the city, accompanied by two minor stations on Rigolets Pass along the Atchafalaya River (figure 2.1). New laws were put into place that also gave the resident physician at the Mississippi station supervisory powers over the installation of a new quarantine system.[15]

The problem with which the new Board had to grapple was simple. The repeated outbreaks of yellow fever had raised alarm in almost all neighboring states. The Board's task was not only to keep the disease out of New Orleans, but also to devise a system that would reassure the State's trade partners within the United States. The biggest economic challenge were so-called shotgun quarantines, imposed in an irregular manner by other states to effectively halt all traffic to and from New Orleans in response to what was perceived as an imminent threat due to the lack of control over the disease in the city. As Gillson has described in detail, the mammoth task of the Louisiana Board of Health therefore involved not only designing a functioning, modern, and reliable system of yellow fever prevention, but also earning the trust of national and international trade partners.[16]

Most considerations in the middle of the nineteenth century affected the time of detention given to ships in the quarantine station. Naturally, opinions differed over the appropriate number of days required to exclude the possibility of emerging cases of yellow fever. Ten days seem to have been widely accepted as a minimum. However, as reports of the Board of Health repeatedly show, contemporary understandings of the nature of yellow fever often suggested twenty-one days of regular quarantine for every ship arriving in New Orleans to be a more appropriate, but also a more costly, timespan.[17]

Figure 2.1
Map of quarantine stations at the Mississippi River from 1886.
Source: Matas Medical Library, Tulane University, "Louisiana Board of Health. *Annual Reports of the Louisiana State Board of Health.* New Orleans, 1886–1887," unnumbered plate following p. 40.

Meanwhile, tradesmen and politicians alike campaigned against an increase in the detention time and continued to assemble evidence for the local origin of the fever; a fundamental opposition to the assumption that quarantine would protect the city in any way remained in place. A prominent case was widely discussed in 1863, when it was argued that even a twenty-one-day quarantine of a highly suspicious ship from Cuba could not prevent the eventual outbreak in the city. Taken as evidence by some of the noncontagious nature of the disease, the case showed to others that twenty-one days of quarantine were still not enough for observing every latent infection, which might have developed en route.[18] Even then, some questioned, how could one make sure that the disease's agents were not still nesting in the ship's merchandise?[19] As Humphreys has argued, opinions regarding yellow fever in the southern states can hardly be limited to just two opposing factions. Rather, the nature of yellow fever was defined from a "continuous kaleidoscope of combinations" and even if the transmissibility was widely accepted, "it left undecided the species of quarantine required."[20]

The Civil War brought about a blockade of New Orleans during its Federal occupation between 1862 and 1865. General Benjamin Butler enforced a strict quarantine against any foreign port suspicious of yellow fever, and instructed the thorough cleaning of the city's streets. In these years not a single outbreak occurred, which bolstered the proponents of the general transmissibility of yellow fever and fostered public belief in the possible advantages of rigorous quarantine systems. But then again, neither in 1866 nor in 1867 did quarantine prove to be a successful protection for the city. Both yellow fever and cholera reached post-war New Orleans and, once again, the usefulness and efficiency of quarantine measures came under attack. Both commercial interests and anticontagionist physicians opposed the strict imposition of quarantine for vessels from Central and South America.[21] Furthermore, the state of Texas imposed another quarantine against New Orleans in 1871, thus contributing to a veritable quarantine crisis. Despite its many critics and a persistent lack of success, long quarantine continued to form a major obstacle to trade and commerce and became increasingly seen as a means of commercial warfare.[22]

In an attempt to resolve the conflict that had lasted for decades through scientific investigations, the Senator to Louisiana, William P. Kellogg, authored a House resolution in 1872 authorizing medical officers to determine "whether any system of quarantine [was] likely to be effective in preventing invasions of yellow fever, and if so, what system [would] least interfere with the interests of commerce at said ports."[23] Following on from the resolution and desperate to resolve a conflict detrimental to the health and trade of New Orleans, throughout the 1870s sanitary officers began to investigate mechanical and chemical solutions to overcome Louisiana's unreliable and costly quarantine system.

Yellow Fever and Carbolic Acid

Humphreys has identified the formation of this new disinfection strategy as a consolidation of the etiology of yellow fever as a communicable disease.[24] By the 1870s most physicians seemed to agree that the fever was caused by some kind of germ, a "poison," often understood to be a living microorganism. The physicians and members of the Board "accordingly set out to destroy that germ with chemicals, and to deny it a hospitable environment through municipal sanitation."[25] Purification through gases and chemical solutions heralded a new scientific age in the prevention of yellow fever.

A report to the American Public Health Association in Boston, delivered by New Orleans physician Charles B. White in 1876, explained the minute details of the new disinfection practice. In an unparalleled attempt to free New Orleans from the scourge of yellow fever, dwellings, streets, and sidewalks throughout the city were disinfected. While the "poisonous cause" of the disease was considered to be not gaseous in nature, still the pathogen ("animal or vegetable") was understood to be attached to the soil, the walls, and surfaces in general.[26] It appeared evident that the agent preexisted signs of sickness. And it was considered to spread in close proximity to places in which cases of the fever had previously been observed. Such dwellings and structures were to be disinfected by "sprinkling floors with carbolic acid."[27] The chemical solution that was used across the city was made of crude carbolic acid and should, according to Perry, contain at least 18 to 24 percent acid.[28] This was at times mixed with cresylic acid, a substance derived from tar oils but that was considered even more unpleasant to residents, even though it was thought to be more deleterious to the yellow fever agents. For White, writing in the 1870s, the so-called "atmospheric disinfection" with gaseous substance was widely useless. For an effective application, the disinfecting substance required direct contact with the poison itself or at least with the surfaces that either produced or retained the cause of yellow fever.[29]

Work at the Louisiana Quarantine Station was accordingly dramatically transformed. In White's opinion, the practice of detention caused not only considerable commercial harm; the combination of long detention with the climatic conditions of New Orleans in summertime was leading to the transformation of ships into breeding grounds for the poison of yellow fever.[30] Detention therefore needed to be urgently replaced by a rapid system of disinfection. From 1867 in New Orleans and from 1870 in the Upper Quarantine Station down the Mississippi River, carbolic acid was first introduced to support the washing of ships, their bilges, and the surfaces of cargo. At the same time, a simple sulphur-based fumigation was carried out to cleanse the air within ships, and to reach into the gaps and narrow spaces that could not easily be washed or sprinkled. To this end, blocks of sulphur were placed in small metal pots, sprinkled with alcohol, and burned in the holds of the cargo ships.[31]

However, at that time, these fumigation measures were far from central to quarantine operations. Looking back from 1904, after the vector of yellow fever had been identified, the author of the Health Board report described this as a critical oversight of the Board's work in the 1860s: "But the traditions of the station would seem to indicate that, by some of those in charge, this measure, the only one by which mosquitoes are destroyed, was regarded rather as an accessory part of the process and was often ridiculed because of its inability to penetrate masses of cargo."[32] To sanitary officers in the 1860s, washing with carbolic acid seemed to take priority. The acid was understood to be an aggressive means of cleaning, trusted with the thorough destruction of possible sources of infection. Yet it was also known to easily cause damage to sensitive freight such as tobacco and sugar. Although less toxic toward possible sources of infection, sulphuric acid seemed to be harmless to merchandise, and its vaporization allowed it to cover entire ships in quick succession.

The Introduction of Sulphur

Growing interest in sulphur in the early 1870s was not purely driven by its chemical qualities, but also strongly supported by its increasing availability, as well as by a drastic fall in its price. In 1869, the first sulphur deposit in the United States was discovered in Calcasieu Parish, Louisiana. Underneath a layer of limestone, a "typical caprock of a sulphur-bearing saltdome" was discovered by the French engineer Antoine Granet.[33] But while sulphur deposits were henceforth regularly uncovered by oil prospectors, over the next few years it remained considerably costly to unearth the "yellow magic" hidden under deep layers of rock. For over a decade various attempts by the Louisiana Petroleum Company failed to drive a high-yield shaft down into the sulphurous pockets.[34] Still, sulphur prices kept coming down, and the abundance of the mineral in the Mississippi area contributed to its reinvention as a substance fundamental to more scientific methods of disinfection.

One of Lousiana's Board of Health sanitary officers, Dr. A. W. Perry, indicated in 1873 that some preliminary tests had been proven successful and he therefore suggested the use of sulphur to build a more sophisticated system of disinfection. In an address to the American Public Health Association, he demonstrated how Louisiana could overcome quarantine as a system of involuntary detention. Enforced and abandoned "already two or three times at New York, Philadelphia and New Orleans," Perry argued, quarantine might perhaps be useful in preventing the importation of diseases from foreign territories, but only if rigorous detention times of over fifty days were applied. However, he argued, as a result of the current system, yellow fever as well as cholera had indeed been imported in various ports: "all these failures have occurred because the quarantine was not based on sound principles."[35] In all cases, no sick person was on board of the vessels, but, he observed, persons involved in the disembarkation of cargo were the first to be affected. One conclusion drawn from this was the possible need for a much longer detention time, and suggestions were made that periods of more than fifty days were desirable. Yet time, Perry argued, reflecting a wider capitalist mentality, was the one element in the quarantine system that was "most oppressive to commerce, the most costly, and at the same time, the least effective."[36] To actually simultaneously fight diseases and reduce detention time, a new method was needed, in which gaseous disinfectants could be used with a special apparatus, whose principles Perry went on to describe.

The new apparatus consisted of one or more force-blowing machines, put in motion by steam power. These were connected with a furnace for generating sulphurous acid gas. By the action of the blowing machines, air charged with the disinfectant was forced through flexible pipes into every compartment of the vessel until it was filled with a saturated atmosphere. The entire apparatus was to be placed on a small steam tug, which would then be moved alongside the vessel requiring disinfection. The air forced into one part of the hold was hence diffused everywhere, and penetrated every crack and crevice by virtue of its elasticity and diffusiveness.

In 1873, Perry thus laid out for the first time the particularities of the system that would eventually be built in the Mississippi Quarantine Station by 1880 and then developed into the Olliphant and later the Clayton machine. The system seemed easily applicable, as steam vessels already provided the conditions to make this procedure highly effective. Most ships had a piping system already installed, with which smoke from an engine fire was distributed and vented equally through and out of the vessel. The very same pipes could be used to apply fumigation. Once a gaseous disinfectant had been injected with the blower, Perry argued, it would saturate every corner of the ship, as gases had been shown to eventually mix and equally saturate across a confined space.[37] The method appeared successful, Charles White reported in 1874:

> The Board during the past year, while insisting on the detention required by law, has made special effort in the direction of disinfection. By the aid of an apparatus planned by Dr. Perry (Quarantine Officer) and put in operations under his supervision, sulphurous acid gas in large quantities was forced into the holds of vessels: carbolic acid being freely used in the bilges, forecastles, etc. As was stated in a special report, either by coincidence, or as cause and effect, on no vessel so treated had a case of yellow fever appeared during its stay in the city.[38]

The 1875 annual report of the Louisiana Board of Health suggested that fumigation practices with large quantities of gas had indeed earned the status of being a reliable resource for the prevention of yellow fever.[39] The development of the first mechanical sulphur furnace raised hopes across the city that a transformation of quarantine legislation was imminent. Changing existing protocols and safeguards was crucial, even perhaps of existential importance, to the city's Chamber of Commerce, which had maintained its resistance against costly quarantines throughout the second half of the nineteenth century. Based on the Board's report, the Chamber issued the following statement:

> Be it resolved, that this Chamber recommends to the present Legislature to grant the Board of Health authority to permit, at its discretion, the passage of vessels from infected ports to the city, after the same have been satisfactorily and thoroughly fumigated and disinfected in lieu of the prescribed time of detention called for under the existing quarantine law.[40]

The Chamber asked for quarantine laws to be amended in order to integrate the new understandings of yellow fever's cause, as this had been proven by the practice of disinfection both at the quarantine station and in the city itself.

In 1875, mounting pressure from experts, commercial interests, and federal politics finally transformed into legislation.[41] Quarantine practice in Louisiana came to be redefined in terms of disinfection. This was mostly enacted through fumigation with sulphuric acid and then washing with carbolic acid. If applied in good measure, even ships originating from heavily infected ports could now sail into New Orleans without any prescribed time of detention. In White's words: "detention at the Quarantine has . . . for the first time been reduced to merely that required for disinfection."[42]

Experimental Disinfection

The experimental undertakings in Louisiana did not go unnoticed by the medical and scientific community in the rest of the United States. If indeed fumigation with sulphuric acid did achieve what it promised at the Mississippi Quarantine Station, it was certainly thought to be of much wider implication for public health both at sea and on land. Indeed, the Louisiana Board of Health's interest in substances, whose disinfecting qualities could be scientifically verified, was embedded in a much larger experimental system. Throughout the 1870s and 1880s, a veritable scientific investigation into the development of reliable disinfectants and the exact conditions (chemical composition, density, time of exposure) of their "germicidity" developed.

In 1884, the American Public Health Association appointed a committee to examine the subject of antiseptics, germicides, and disinfectants in their specific relationship to preventive medicine and sanitation.[43] Drs. George M. Sternberg (Surgeon to the U.S. Army), Charles Smart (National Board of Health), and George Rohe (Professor for Hygiene at Johns Hopkins University) investigated by means of experiments the germicidal value of various substances that had been so far employed as disinfectants in homes, streets, and ports. The question was, what were the germicidal capacities of carbolic acid and sulphur, and how did these substances interact with a range of common materials to be found in the holds of vessels and in private homes? Systematic research began in the biological laboratory of Johns Hopkins University in 1884 with the exposure of a variety of microorganisms to increasing amounts of the two gaseous substances.

The researchers first aimed to verify the effectiveness of carbolic acid, which in liquid form was widely understood to have excellent disinfecting capacities and had long been used both in laboratories as well as in quarantine and increasingly also in private households. Historically, carbolic acid was attributed to Friedlieb Ferdinand Runge, who had discovered the antiseptic in 1834 in its distillate form of coal-tar. In 1876 the substance was brought to prominence by Joseph Lister who introduced it into antiseptic surgery.[44] Carbolic acid was well known across the globe for its ability to halt putrefaction, but its capacity for killing disease agents was subject to controversy. Progress in determining the concentration in which it was indeed "successful in destroying the virus" was made in experiments on vaccines in Britain.[45] There, John Dougall carried out inoculation with "infectious matter" that had been exposed to carbolic acid, and his research demonstrated that such cultures did not lead to any symptoms or illness; the agent was therefore considered neutralized.

As American researchers quickly discovered, carbolic acid's germicidal capacity diminished when it came to its gaseous vapours. The atomization of 5 percent carbolic acid failed to achieve the same disinfecting power compared to washing of surfaces with the original liquid. Furthermore, bacteria of various diseases in exposed liquids were destroyed by vapors of 7.5 percent carbolic acid, while bacteria on clothes could withstand even an atomized 12.5 percent solution, when the clothes were damp. In summary, the report confirmed the well-known disinfecting properties of a 5 percent liquid solution of carbolic acid against almost all bacteria and fungi.[46] But the large proportions necessary for acquiring true disinfecting qualities led the report to conclude that a vaporized form of carbolic acid should not be assumed to be effective in any way.

As was widely assumed, the researchers could show that the case was different for sulphuric acid. Sternberg referred to the French chemist Émile-Arthur Vallin and his authoritative treatise on disinfectants to argue that sulphur dioxide was one of the most promising substances for the future of practical disinfection.[47] Back in 1882, Vallin had countered Robert Koch's scepticism on the efficacy on sulphur dioxide by conducting his own experiments with the substance, which left him satisfied that it was "a powerful disinfectant."[48] Indeed, Sternberg argued, "no gaseous disinfectant known is more extensively used, or has a higher place in the confidence of leading sanitary authorities at the present day."[49] Still, a critical revision of literature and the experiments conducted by other bacteriologists led Sternberg to a much more critical conclusion. While he believed that a range of experiments demonstrated that exposure to the gas killed various live bacteria, he also showed that bacteria in fabrics or other porous materials did regularly survive gasing at a wide range of exposures and densities of the gas. As a result, he queried whether "sulphuric acid" was a reliable agent only when it came to killing microorganisms suspended in atmosphere and nesting directly on nonporous surfaces. Surely, Sternberg concluded his report, the existing methods of washing with carbolic or other acids were much more reliable.[50]

Yet laboratory conditions could not easily be translated into the situation on the ground. To verify the usefulness—if indeed there was any—of disinfection with gaseous substances, J. H. Raymond, Professor of Physiology and Sanitary Science in Long Island College Hospital and Health Commissioner of the City of Brooklyn, devised an experimental setup in a common living quarter in Brooklyn, New York. The experiment was performed in 1884 at the request of the U.S. Commissioner of Health so as to establish if "sulphurous acid gas" could be used as a reliable disinfectant in the country's main gateway to the East.[51] "Sulphurous acid gas" was experimentally applied by Raymond to a closed room, which had been equipped with various prepared infectious materials from Sternberg's laboratory at Johns Hopkins. After all cracks and crevices were closed with cotton, the materials were distributed in the room: small pieces of blankets were soaked in the blood of a rabbit that was killed while being affected by septicemia; other pieces where soaked in the blood of another rabbit, which was infected with anthrax. Some of the blanket pieces were folded and some were hidden under or within furniture in the room. Furthermore, a container with live vaccines in liquid solution was placed in the room. In a large coal scuttle filled with wet ashes, sulphur was moistened with alcohol and subsequently lit, and the room was fumigated for ten hours. In a follow-up experiment, blood as well as organic matter taken from the linen was used to inoculate healthy rabbits. Both septicemia and anthrax prevailed, and most of the subsequently infected rabbits died. However, further experimentation with the same setup would show that the particular gas was able to kill the live vaccines that had been directly exposed in agar.[52] The questions remained: which was the right solution, what was the time required for effective germicide, and what were the appropriate means of application in different spatial contexts? This was not an isolated experiment. Indeed the 1880s saw the onset of an extensive national and international debate over the germicidal properties of a long list of chemicals.

On the other side of the Atlantic, both Pasteur and Koch had investigated the capacities of sulphur-based gases. In Germany, Dr. G. Wolffhuegel experimented, in 1880, in Koch's laboratory with sulphurous acid in closed rooms to investigate the density required to kill insects.[53] In France, in 1879, Pasteur experimented with sulphuric acid as a reliable method to suspend the development and virulence of the bacteria causing fowl cholera.[54] The 1880s hence witnessed the emergence of an international landscape of experimental approaches to the disinfecting capacities of sulphur-based gases whose application as a means of disease prevention appeared promising in hospitals, warehouses, private homes, and sewage systems.

From the Holt System to the Clayton Apparatus

In parallel to the scientific validation of the disinfecting qualities of "sulphuric acid gas," New Orleans undertook in the 1880s a thorough modernization of its quarantine stations. This was both motivated by the success of the experiments developed locally and encouraged by the international scientific ratification of the germicidity of SO_2. In 1884, the state of Louisiana was invited to an interstate quarantine conference aimed at resolving tensions and dissolving the atmosphere of mutual suspicion between the southern American states. A notable decision of the conference was the abolition of "non-intercourse"—a term referring the absolute ban on trade and traffic from Louisiana justified by epidemic threats. On the one hand, the decision was supported by the increasing scientific accountability of disinfection methods. On the other hand, the proposition of a new reliable system of maritime sanitation was high on the agenda.

The new vision was brought forward by a prolific Quarantine Officer, Joseph Holt, who became the director of the Louisiana Board of Health in 1884. A few decades later, he was acknowledged as the architect of the new scientific mode of quarantine that would finally overcome what Holt framed as "barbarous" detention.[55] The brainchild of a public political figure, Holt's legacy was not built solely around the mechanical improvement of the quarantine stations protecting New Orleans. It also involved raising questions of quarantine and disinfection in countless public lectures, speeches, and published pamphlets. Holt believed quarantine and commerce to be ultimately aligned in their interest. Both public health and commerce would eventually profit from a quarantine system, if this were solely erected upon those practices of disinfection that had been verified by rigorous scientific observations. Quarantine, Holt argued emphatically, should be limited to a "sharp, narrow channel, obstructive to importation of pestilence, but open as a highway to commerce."[56]

When Holt took over, the mechanical furnace originally designed by Perry and slightly modified by his successor Joseph Jones had been in place for almost a decade. Six years before Holt took office, a large outbreak of yellow fever in 1878 had put to bed illusions that the new furnace would be a rapid as well as a safe method for incoming vessels. Furthermore, the Board's severe lack of funds had prevented it from installing additional furnaces to arrive at the required capacity to deal with the growing commercial traffic across the Mississippi.[57] Instead, the Louisiana quarantine system became embroiled in national controversies regarding initiatives for a national quarantine system. Due to the lack of a functioning system of disinfection and fumigation, and as the furnace was no longer believed to provide satisfactory results, in 1884 (just before Holt took up his position in the Louisiana Boards of Health) New Orleans saw the brief return of involuntary detention times of up to forty days for all ships arriving from suspicious ports. Given the public protest against the return of a costly, unreliable, and outdated practice, Holt defined his mission as the abolition of detention time in exchange for the implementation of a fumigation system based on sound scientific principles.[58]

To reconcile commercial interest, quarantine requirements, and public health, Holt presented his new design of "Maritime Sanitation." The new director of the Board declared that "time was no factor in clearing a ship of danger; but that decisive action in the immediate cleansing of a vessel and of all that she carried aboard, offered the only rational hope of defence."[59] Attacking the ship's uncleanliness using the force of local fire departments instead of leaving it to broil under the July sun was, according to Holt, not just the only rational and humane thing to do, but was also proven to be practicable and scientific. As Gillson argues, "Holt's celebrated system of maritime sanitation introduced in 1885, was basically pragmatic in its application."[60] His new system rested primarily on two quarantine stations and two supplementary points of inspection along the Mississippi River.

The regulations required all vessels to be inspected in Port Eads, 110 miles away from New Orleans. Here logs were controlled and a sanitary officer would take a sworn statement from the captain regarding any incident on the vessel's journey. All ships were immediately sent to the Upper Mississippi Quarantine Station (figure 2.2).[61] If found clean and arriving from a healthy port, they were sent along to New Orleans. If, on the other hand, a ship was found to have arrived from an infected port, its passengers and crew had to disembark, while the ship was washed in a solution of bichloride of mercury, its holds were fumigated, and any luggage was disinfected. To this end the Board had invested in a tugboat (figure 2.3), which was equipped with a "complete outfit for generating and applying germicidal gas, for displacement of the entire atmosphere within the ship."[62] Eighteen furnaces delivered the rapid combustion of sulphur. The gas was blown through pipes so as to enter the holds of the ship (figure 2.4). The gas then permeated every kind of cargo, even excelling at the highest test: the disinfection of coffee. Everything that could be removed from cabins, such as

(PLATE 2.)

View of disinfecting wharf, showing tug fumigating vessel; elevated tank containing 8000 gallons of bi-chloride of mercury solution, 3 leads of hose from tank to ship. Gangway leading to building containing super-heating chamber.

Figure 2.2

Illustration of the Upper Mississippi Quarantine Station, with a moored barque, attached to a tugboat with a fumigation apparatus.

Source: Matas Medical Library, Tulane University, "Louisiana Board of Health. *Annual Reports of the Louisiana State Board of Health.* New Orleans, 1886–1887," unnumbered plate following p. 40.

linen, luggage, curtains, and other fabrics, was placed in the large super-heating chamber at the station, which sterilized large amounts of goods in quick succession. Infected ships carrying human cases would undergo the same treatment, but would be directed to the lower quarantine station where a hospital and isolation stations were equipped to deal with the patients.

The system remained in place for the subsequent years and yielded to astonishing success. Dr. Kennedy, a physician from New Orleans, wrote in 1885 that the city had finally removed quarantine from the "dark shadows of the ignorant past," and placed into the "bright sunlight of science."[63] Holt took pride in having revolutionized the quarantine system and believed that he had contributed to a vast improvement of morale in the city as an effect of the kind of safety his system promised.[64] Indeed, not a single case of yellow fever was imported along the Mississippi River until 1892, raising interest into Holt's Maritime Sanitation system on a national and international level. In 1888, the Marine Hospital Service dispatched the emerging bacteriologist J. J. Kinyoun to inspect the system and to report on whether it was feasible to install it nationwide.[65] Kinyoun found the results to be impressive but also noted some mechanical failures, which the Board began to correct immediately in 1889. Subsequently, Surgeon General Hamilton ordered all national quarantine stations to secure the necessary apparatus and to build heating chambers, as well as fumigating devices, as devised by Holt.

In April 1890, the newly appointed Board president Samuel R. Olliphant gave an address to the public to outline the new policy and plans of the Board of Health.[66] While he emphasized the pivotal position of New Orleans, shouldering the responsibility to safeguard the commercial interests for the entire United States, he also praised the achievements of what was now known as Holt's "system of 'maritime sanitation.'"[67] The report of the Board's quarantine committee was then read by its newest chairman, none other than Thomas A. Clayton. After he had been appointed by Govenor Nicholls in early April 1890, he had joined the quarantine committee to represent farmers' economic interests and to integrate their demands into the health policy of the state.[68] In his report, Clayton described his own first encounter with the sophisticated and spectacular system that was now in place at the quarantine station. A Portuguese barque, *Maria*, had arrived on April 17, 1890, from Rio de Janeiro. The barque was towed to the

quarantine station, linen and clothing were removed and brought to the
hot air cylinders, while surfaces had been washed with bichloride of mercury. Only then could the new fumigation system be brought into action.
In this case, the ship was brought alongside the wharf, where a fumigation apparatus on rails rolled adjacent to the vessel to then commence two
hours of intensive fumigation with sulphuric gas. Clayton pointed out that
the capacity of the stationary apparatus was limited. Citing the growing
demand, he motioned for a further tugboat to be equipped with an apparatus for fumigation operations on anchored vessels midstream.[69] Clayton
declared publicly his trust in the system to effectively prevent the introduction of any disease while providing for an almost seamless transfer of goods.

In 1890, Olliphant could proudly present the Board's successful abolition of the "old brutal and haphazard quarantine methods": Holt's system
of Maritime Sanitation had proved to be scientifically reliable and economically sustainable.[70] In June of the same year, a triumphant Olliphant

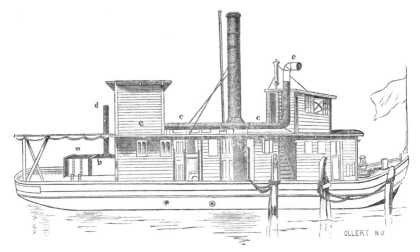

(PLATE 1.) TUGBOAT WITH FUMIGATING APPARATUS.

a. Furnace. *b.* Reservoir for reception of gas. *c.* Discharge pipe, conveying gas to ship's hold. *d.* Escape
pipe for gas when fan is at rest and sulphur is burning; closed by a valve when fan is in motion. *e.*
House protecting from weather the machinery for driving fan and containing accelerating gearing.

Figure 2.3
Illustration of the New Orleans tugboat with a fumigation apparatus installed.
Source: Matas Medical Library, Tulane University, "Louisiana Board of Health. *Annual
Reports of the Louisiana State Board of Health.* New Orleans, 1886–1887," unnumbered
plate following p. 40.

invited officials for a formal visit of the quarantine station operations, so the govenor and a group of selected physicians could "see for themselves the effective system of disinfection and fumigation in operation at the mouth of the river."[71] Five years later, Olliphant could also point out that many faults and problems with the original system had been fixed; the danger of damaging cargo or merchandise was practically nil. Clothing and bedding of passengers continued to be treated in heating chambers. Olliphant happily presented his own modification of the apparatus aimed at overcoming its previous flaws. Due to limits in the combustion system, the saturation of the SO_2 gas had previously not exceeded 4 to 8 percent in the process of displacing the oxygen in the holds. His modified furnace was hence equipped with a second pipe to extract the oxygen used to ignite the sulphur directly from the vessel's hold. It thus effectively exchanged the air inside the ship directly with the germicidal gas, and the resulting circulation could eventually deliver highly effective solutions saturated to up to 15 or 20 percent.[72]

Olliphant presented letters from doctors and researchers supporting the validity of his claims and promised to continue the "movement," brought into motion by Joseph Holt, of setting quarantine on scientific basis. When cholera reared its head again in 1892, Olliphant got in touch with over sixty quarantine stations around the world, asking for the specifics of their disinfection methods and the substances used. After reviewing the answers, he concluded that nothing could be done to improve the running service at New Orleans, that it was clearly the best system available worldwide.[73] In January 1893, the apparatus was patented as an improved Fumigation Apparatus.[74] The patent was issued in the name of both Samuel R. Olliphant and Thomas A. Clayton. In their application, the inventors defined the objective to provide a fumigation apparatus, which would produce a "gas of ample and certain strength so that it would be absolutely sure to kill all disease-causing germs."[75]

These efforts were supported by the dramatic decrease in the cost of sulphur, which became an increasing focus of the commercial landscape in Louisiana. This was the result of a breakthrough delivered in 1891 by the engineer Herman Frasch. Frasch had patented a new sulphur extraction method, which he had tested on his own property, located right by the unsuccessful mining operations of the Louisiana Petroleum Company. His method, called until today the Frasch process, would eventually

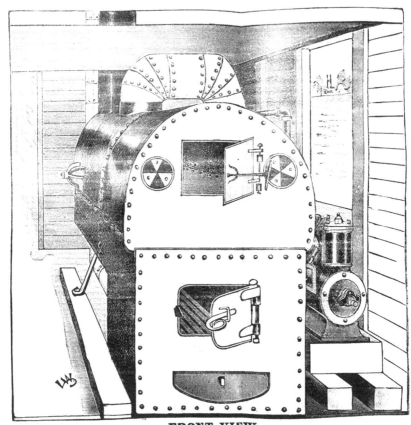

FRONT VIEW.
Of Improved Furnace for the production of Sulphur Dioxide Gas, showing sulphur furnace with fire box underneath, and curved pipe carrying the gas into reservoir.

Figure 2.4
Illustration of the new and improved furnace inside the New Orleans tugboat, used to produce sulphur dioxide gas.
Source: Matas Medical Library, Tulane University, "Louisiana Board of Health. *Annual Reports of the Louisiana State Board of Health.* New Orleans, 1886–1887," unnumbered plate following p. 40.

revolutionize the industry and cut to a fraction of the original price the cost of sulphur-mining in areas where it was not found close to the surface. Frasch exploited the fact that sulphur melted at 240°F, and after drilling a mining hole with common petroleum extraction equipment, he inserted a number of pipes, some of which were perforated. Pumping large amounts of boiling water into the sulphur pockets melted the sulphur, and through a different pipe and the introduction of pressurized air, the liquidified sulphur could be extracted onto the surface. There, it was simply laid out to dry, before the yellow rocks were ready for shipping at the Mississippi port close to New Orleans, which carries until today the name Port Sulphur.[76]

In 1893 a model of the quarantine system of Louisiana was exhibited at the World's Columbian Exposition in Chicago. The accompanying leaflet boasted: "The improved system, as operated by the Louisiana State Board of Health, aims at the complete destruction of all germs of disease, wherever existing, with the least injury to private property and the minimum of detention to vessels and passengers."[77] Its inventors presented the model as one that could achieve the purification of the vessel, as well as the full disinfection of anything carried on board. The brochure emphasized the blowing mechanism made by Olliphant, which enabled a saturation of over 18 percent sulphur dioxide in the ship's atmosphere. It also placed the apparatus at the center of the disinfecting operations in Louisiana, including the washing with bichloride of mercury and the steaming of fabrics and personal belonging.

The System of Maritime Sanitation was further characterized by the location of inspection and quarantine stations, which allowed for the calibration of measures and preventive procedures according to the kind of threat that separated infected and "foul" vessels from those arriving from infected ports with clean records. It was within this system of locations, practices, and calibrations that a fumigation apparatus began to take on an essential role: replacing unreliable and dated detention times with modernized and specified interventions, as it is "against the *germ* alone that modern sanitary science urges the warfare of quarantine."[78] With a celebratory note, the model of Louisiana's sanitation system in the Exposition announced that "through the operation of this complete system of quarantine work, the Board has succeeded in keeping out foreign pestilence, and at the same time has fostered the commerce of its State and section to a gratifying extent."[79]

As already mentioned, the new apparatus was patented under the name of Olliphant, as well as Thomas A. Clayton. But Clayton was destined to further develop this new, revolutionary method of maritime fumigation, leading to a machine that would be employed not only in the American South but eventually across the globe. Born in Banff, Scotland, in May 1852, Clayton had arrived in Louisiana in 1878, were he conducted various businesses for twenty-two years. He first purchased a plantation in the St. Landry parish and became a successful cotton farmer, despite his lack of experience in this field. In 1882, Clayton took charge of the Farmers' National Commission office in New Orleans and became an outspoken proponent of the alliance movement: the National Farmers' Alliance and Industrial Union ventured to became a stronghold of farmers' economic cooperation across the American South and sustained pressure on the state and the nation in terms of transportation, valuation, land ownership, and access to markets.[80] Clayton had organized the alliance lodge in the St. Landry parish, near New Orleans, and became subsequently a secretary of the executive committee of the state alliance and joined later the populist cause of the people's party. As a manufacturer, he was known across Louisiana's society to possess "courteous and agreeable manners."[81] Furthermore, as the chief purchaser, he acted as a wholesale agent for the acquisition and sale of farm products for the state's farmers. After Clayton was appointed by Govenor Nicholls to the Board of Health in April 1890, various entries in the records suggest a strong bond between him and Olliphant across the various businesses of the Board.[82] But already in 1891, Clayton saw himself quickly embroiled in a scandal, as he stood accused of having tampered with health certificates to slaughterhouses, rather then following the rigid routines of sanitary inspections. Despite his claims to the contrary, Clayton immediately resigned.[83] His next appearance in the city's newspaper came when his newly opened garbage incineration plant did not work as promised.[84] Due to low capacity of garbage—so Clayton defended his failing business—the plant did not work properly, thus sending the smoke down to the city's center instead of over to the Mississippi Delta.

Clayton, an avid merchant, seemed to have moved into the business of producing fertilizers and chemicals by 1895.[85] It might have been here that he discovered the astonishing fire-extinguishing capacities of the sulphuric gas used in the Olliphant apparatus installed at the quarantine station. Perhaps motivated by the falling price of sulphur at the time, he investigated further. At the time, chemical solutions to the task of fire extinguishing usually relied on carbolic acid, a good mixture usually requiring about 15 to 20 percent density to have the desired effect on blazing flames. By contrast, with sulphuric acid, as Clayton found out, a density of only 5 percent had similar effects. In the last years of the nineteenth century, Clayton began to develop his very own apparatus that would provide both functions in one machine: an apparatus that would be usable both as a fumigating as well as a fire-extinguishing device. He thus registered his patent on June 7, 1899, filed as "Method of and Apparatus for Fumigating and Extinguishing Fires in closed Compartments."

Clayton and Olliphant appear to have pursued common commercial interests already from 1892, seeking to capitalize on the success of their apparatus. As the *Washington Post* reported in December 1897, Olliphant and Clayton had pursued the registration of their patent not under entirely fair circumstances. Holt's original invention had been made freely available to the public, intended explicitly for the benefit of humankind. After they had appropriated the system and registered patents in their name, Olliphant and Clayton approached Walter Wyman, Surgeon General of the Marine Hospital Service, to sell their patent to the US government. Wyman rejected the offer "on advice of patent authorities and law officers" and added: "Claim entirely untenable."[86] However, the grand jury that was tasked by the Board of Health to dispute the patent saw itself overwhelmed by yet another outbreak of yellow fever in 1897.[87] As the Board of Health resigned, the case was eventually dismissed.

The Clayton apparatus advanced into global trade despite its apparent failure to protect New Orleans in 1897. A reason could perhaps be found in the design of the 1899 patent by Clayton, which described an advanced apparatus, smaller and more powerful than the machine he had built previously in collaboration with Olliphant in 1894 (figure 2.5).[88] The new machine did fit easily on a bed plate and could be transported without too much effort. It consisted of a generating furnace, in which sulphur was burned, a cooling system to regulate the temperature of the sulphurous acid gas, and a power blower through which the gas could be expressed with high pressure. The furnace was a plain insulated chamber, divided horizontally by a wire netting. A door through which dry sulphur pieces could be introduced was placed at the front, and two pipes at the rear of the chamber. In Clayton's new design, an additional system of pipes allowed

the fumes to circulate within the machine so as to achieve a rapid enrich-
ment of the vapors, which could afterward be cut off through a by-pass.
Through this system, the burning sulphur was supplied with large quantities of air by an induced draught. The high temperatures of regularly about
1800°F achieved in the furnace enriched the SO_2 with SO_3 and other sulphur oxides, before it was cooled down. Compared to the burning of sulphur in open fires, where SO_2 remained the almost exclusive compound,
the burning in the furnace achieved the production of nearly a sixty-fold
amount of SO_3. This, it was assumed, drastically raised the toxic capacities
of the gas and contributed also to its excellent fire-extinguishing qualities.
Another advantage of the resulting gas was its cloudiness, which was again
a result of SO_3, which made it visible to those operating it and thus obviated
accidents. Lastly, a second circuit with flowing water allowed the gas to cool
down toward a temperature that had been shown to increase its germicidal
as well as it fire-extinguishing capacities.[89]

A central idea embodied in this, as well as in the previous model developed in collaboration with Olliphant, was to establish a continuously circulating flow of air and gas. The air, extracted out of the holds of the ship, was
used to raise the temperature in the furnace to thus continuously increase
the amount of SO_2 in the mixture. A problem with the old Olliphant system, however, had been that, already at about 5 percent of gas, the fire in
the furnace would start to diminish. In other words, at about 5 percent
the mixture produced in the furnace had acquired such fire-extinguishing
capacities that they became detrimental to the continuous burning of sulphur. To solve this problem, Clayton developed his design further, aiming
to achieve two objectives. Clayton's patent states: "Preferably I contemplate
the generation of gases or vapours in a furnace the intake and outlet of
which are connected with a closed compartment and through which and
the compartment a forced circulation is maintained, so that the gases will
be rapidly enriched to the desired point, whereupon I cut the furnace out
of the circuit by means of a by-pass."[90] Clayton thus added a valve that
registered whenever the returning air from the compartment reached 3 percent, so that a connection between the return pipe and the furnace would
be closed, and another valve would be opened to induce fresh air into the
system to keep the fire in the furnace burning. With the same mechanism
the circulation was shut off while the furnace continued to pump gas into

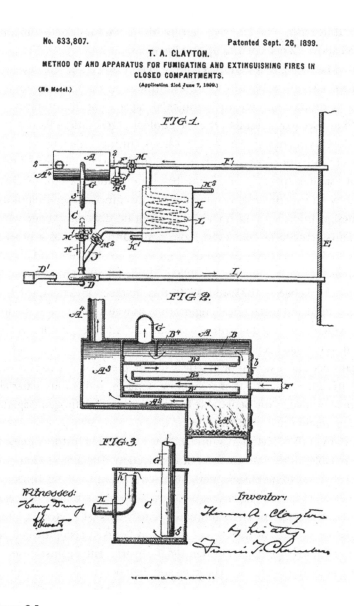

Figure 2.5
Diagram of the Clayton apparatus.
Source: US Patent Registry.

the ship, thus raising the effective density inside the fumigated compartment well above 12 percent and up to 15 percent.

It would be wrong to assign the chain of events that led to the development of the Clayton apparatus to a breakthrough in the scientific understanding of disinfection and the germicidal capacity of disinfectants, or indeed to just see it as a result of the bacteriological revolution. Rather, the birth of the Clayton machine in New Orleans is a history driven by the abolishment of "barbarous" detention times and the experimental reinvention of the old sanitary practice of sulphur-based fumigation. A process characterized by an experimental employment of machines and contraptions, whose design informed and impacted a scientific discourse about what disinfectants actually do. But even more importantly, this was a process that involved the successive design of furnaces, power blowers, and cut-off valves aimed at producing a gas capable of killing all germs in all possible circumstances. Between 1875 and 1900, this involved an entangled history of chemical and mechanical engineering, as well as of medical theory, through which maritime quarantine was radically redefined, if not actually abolished. What had been a vaguely defined barrier to diseases transmitted by humans became a system of practices set up to eradicate germs. It was not human carriers, but goods, merchandise, surfaces, and porous materials that were identified as the penetrable hiding grounds of pathogenic agents in the experimental setups of New Orleans.

The origin of the Clayton machine in New Orleans is a history of developing a reliable industrious method of disinfection. The aim of Holt and his successors was to mobilize a method proved to destroy the pathogenic capacity of bacteria into the large-scale conditions of maritime trade. Not only did this practice promise ultimate security against bacteria hiding on the porous surfaces of the vessels' walls and floors, it also provided an effective—although perhaps unintentional—instrument against insects. Yet, with a predominant focus on bacteria and their possible hiding grounds, the experimental systems erected in New Orleans sidelined rodents and, in particular rats, which had in the meantime moved into the focus of comparable systems on the other side of the Atlantic.

3 Rat Destruction in Istanbul and Hamburg

Between the nineteenth and the twentieth century, with rats becoming increasingly a subject of epidemiological attention as carriers and spreaders of plague, fumigation technologies began to be developed in Europe with the explicit purpose of eradicating this particular animal in the holds of steamships. In this chapter we will explore the emergence of these deratization technologies as it unfolded in two distinct locations: the capital of the Ottoman Empire, Istanbul, and the most important harbor of recently unified Germany, Hamburg. These examples make for an unusual sulphuric utopia, as in both countries the chemical compounds chosen to yield best deratization results were indeed carbon based. Nevertheless, we take these sites of experimental fumigation as integral parts of the global history told in this book under the rubric of a hygienic sulphuric vision, as they encapsulate similar principles and comparable experimental systems developed as competitors to sulphurization. Their aim was to achieve the one and same goal: a complete and combined disinfestation and disinfection while causing only minor obstruction to sea trade and minimal damage to vessels, goods, and merchandise. There is, however, another reason why examining the development of maritime fumigation in these two locations is important. Misleadingly, these two sites are often taken as two opposite ends of the spectrum of governmental and scientific development in Europe. On the one hand, the Ottoman Empire is commonly seen as a moribund state, the "sick man of Europe" stagnating in its inability to adapt to the new age of science, technology, and modern statecraft. On the other hand, unified Germany under the Kaiser is commonly seen as the most rapidly industrializing state in Europe, with scientific development tightly tied to industrial expansion and the raison d'état. And yet, as historians like Miri Shefer-Mossensohn, Amit Bein, and Alper Yalçinkaya have recently shown, the late Ottoman Empire was a hotbed of scientific debate and innovation.[1] As Zeynep Devrim Gürsel has argued, this was particularly pronounced in the realm of medicine.[2] At the same time, as late as 1892, Hamburg remained a seat of scientific and governmental resistance, indeed one may properly say reaction, to contemporary bacteriological breakthroughs.[3] The choice of these two sites for the study of the development of antirat maritime fumigation is thus driven by the need to go beyond a comparison based on historical-technological stereotypes (the progress-oriented West versus the tradition-stuck East), and toward a crossed history of technological and public health development that does justice to the political and political economic entanglement of maritime sanitation on the international stage. This will allow us to glimpse how, at the turn of the century, a distinct hygienic utopia, centered on the eradication of rats, arose across Europe, leading in turn to the emergence of chemical fumigation technologies—technoscientific apparatuses that aimed to both protect respective nations or empires from the global march of plague and to regulate maritime trade in ways that were beneficial to respective financial and geopolitical interests.

Rats as an Epidemiological Problem

Today plague is so closely associated, indeed identified, in popular imagination with rats, that it is easy to forget that before the 1870s the rat was generally considered to be a nuisance but not the carrier or spreader of diseases. Catching and destroying rats was of course part of elaborate processes, involving ferrets, dogs, and skilled rat-catchers in a "multi-species labour of rat-catching."[4] These were processes that, on the one hand, involved public spectacles of "ratting," while, on the other hand, developed notions of rat intelligence and intentionality, as is famously evident in Henry Mayhew's 1850s account of the practice.[5] As Neil Pemberton has noted, "these cultural practices invested rats with a menacing and formidable persona: a species co-existing and co-emerging with civilization, devouring it from within."[6]

This was a kind of "hostility" against rats that, in England at least, was further fueled by the discovery of a displacement of the black rat by a malicious "invader": the brown rat. Pemberton notes how, in tandem with naturalists like Charles Waterton, Mayhew construed the brown rat "as a foreign species, notable for its particularly 'rapacious' appetite: calculating and scheming to destroy habitats and human food sources, while cleverly hiding its booty."[7] But even when discussing its otherwise iconically vicious bite, Mayhew failed to note any infective qualities of this particular rodent. Indeed, the rat was considered to be uniquely able "to 'clean' and preserve itself from contamination by the filth and miasma of the sewer."[8] Pemberton notes that "rather than being correlated with plague, the sewer rat's appetite for putrefying matter saved human inhabitants from 'periodical plagues,' which Rodwell insisted were the 'result of deadly gases arising out of the putrefaction of animal and vegetable matter.'"[9] Until the final decade of the nineteenth century, the rat problem was thus not a problem of infection but rather a problem of boundaries and their transgression, including both the unwarranted nocturnal wanderings of the rat into private and familial spaces, and its "invasive," transnational character.

The first note of the rat's implication in human disease came in the mid-nineteenth century from two distinct sources, both from what at the time were considered to be "remote" parts of Asia. First, in the 1840s, British colonial officers noted that in the Garhwal and Kumaon Himalayan districts of the British Raj a disease known as Mahamari (suspected by British doctors to be plague) was claimed by locals to first appear in rats before striking humans.[10] Three decades later, news of a widespread plague outbreak in the Chinese province of Yunnan also reported similar local beliefs.[11] It was, however, not until the Yunnan-originated plague reached Hong Kong, in 1894, that the rat became the object of systematic scientific study and problematization. By the time the outbreak was in full sway, leading medical colonial officers, like James Lowson, still maintained that the disease first struck rats because the plague gases emanating from the soil first reached the nostrils of the earth-bound animal, and humans only later because of their heads standing higher above ground. However, already in his paper announcing the discovery of the plague bacillus in July 1894, Alexandre Yersin included the rat among the nonhuman suspects of carrying and perhaps spreading the disease.[12]

It was not until 1898 that a direct link was established, this time including the rat's flea *Xenopsylla cheopis*, as a rat–human vector. The theory was formulated by another Pasteurian, Paul-Louis Simond, but it was not immediately endorsed by the international medical community, which was at the time faced by the rapid spread of the third plague pandemic across the globe.[13] It would take another eight years for the rat–flea theory to be fully accepted, and only in 1914 would Bacot and Martin, of the Lister Institute, be able to demonstrate the mechanism through which plague passed from fleas into rats and humans.[14] Yet this did not mean that the rat theory remained shelved and ignored. By contrast, the question of the rat formed the ground for an impressive range of studies, theories, and experiments that still await their historian.[15] In the course of this book, we will encounter several of these, particularly as they relate to maritime trade. It is, however, useful to keep in mind that besides the problematization of seaborne rats, the scientific study and intervention on the particular rodent extended into *terra firma*, where other means of rat elimination besides fumigation were developed and deployed—although, for reasons that will become clear, these were largely unemployable on board of ships.

This is not a book on the history of epidemiological approaches of the rat, nor is it a work on the history of the global war against the particular animal as waged from 1894 onward. Instead, our particular focus is on the aspiration of complete maritime sanitation by means of mechanically produced fumigants. In this story, the rat played an important role as, at one and the same time, the most visible and quantifiable opponent. Rats could not only be retrieved from the holds of ships, but could also be counted in order to demonstrate the efficacy of fumigants in ways that insects or bacteria could not. At the same time, rats escaping the deadly grip of fumigation were equally readily observable by the naked eye, even by the least scientifically versed members of the crew. Rats thus operated as readily available proofs and disproofs of the efficacy of different fumigation methods and technologies. However, as we will see, at the dawn of antiplague fumigation, disinfestation did not form an autonomous target, but only one that mattered in conjunction with disinfection. In other words, killing rats in ships' holds was configured as meaningful only to the extent that the gas employed was also shown to destroy the bacteria carried by rats.

Polis Mytilini and Apéry's Machine

In spite of the occasional epidemic in Persia, Mesopotamia, and Benghazi, Ottoman concerns with plague ran relatively low in the second half of the nineteenth century, when, since its first outbreak in Istanbul in 1831, cholera formed the principle worry of health professionals.[16] In the context of wider modernization reforms leading up to *Tanzimat* (1838), and in light of resurgent plague outbreaks in the region and of international concerns over the impact of cholera across the globe, mounting European pressure on the Sublime Porte led to a series of radical changes in public health. These included the establishment of hospitals and the first Ottoman quarantine regulation, issued as early as 1836.[17] It is in this context that in 1839 the Quarantine Council (*Meclis-i Tahaffuz*, also known as the Conseil Supérieur de Santé) was established in the Ottoman capital.[18] This was "tasked with enforcing quarantine regulations in the Mediterranean region. The Council initially consisted of eight Ottoman members and delegates from nine European states (Austria, Belgium, France, England, Greece, Prussia, Russia, Sardinia, Italy). Sixty-three sanitary agencies dispersed across the Ottoman domains reported to the council."[19] As Nuran Yıldırım explains, it was on the same day as the foundation of the Quarantine Council (June 10, 1839) that the Organizational Regulation for Maritime Arrivals was instituted, coming into effect two months later:

> According to this first regulation, every ship that came to Istanbul was to have a health patent to be delivered to the health control officer by means of a long pole. Patents that were given 30 days after the last case of plague were considered to be clean; those given after 15 days were suspect, while those given before 15 days had passed were considered to be infected. Ships with suspect patents and ships with a 10-day infected patent were to wait in quarantine for 15 days. Their goods and passengers were to be placed in the Kuleli quarantine. Every ship, be it from the Mediterranean or the Black Sea, was to be subjected to interrogation and the captain of the ship was to declare the health conditions on board.[20]

Following Yıldırım, Sultan Abdülmecid invited the inclusion of delegates from foreign embassies in the Quarantine Council as this was believed to foster international cooperation, dispel harmful rumors, and promote expediency and efficiency in the implementation of quarantine measures. With the Ottoman members of the Quarantine Council being a minority, its international status was further enhanced as a result of the 1851 International Health Conference in Paris demanding the Sublime Porte to recognize the Quarantine Council as bearing legal status and to give voting rights to foreign delegates. Delegated with "the authority to prepare laws concerned with epidemics and quarantine," the Quarantine Council issued decisions that had to be implemented by local quarantine officers, who were appointable and dismissible directly by it.[21] This placed consider-able power in the hands of an organization that by 1897 was solidly under foreign control.[22]

With the notable exception of studies of how it affected the pilgrimage to Mecca, Ottoman concerns about and approaches to plague following the global resurgence of the disease since the 1894 outbreak in Hong Kong remain historically underexamined.[23] Nonetheless, all evidence confirms that international outbreaks, particularly those in the Indian subcontinent, were duly noted in leading Ottoman medical journals at the time. More than being distant events, these raised the question of whether a series of short plague outbreaks in Jeddah (May 1897, April 1898, February 1898), the entry port for accessing Mecca, were derived by means of Yemeni pil-grim caravans from what were generally held to be local endemic regions (the high plateaus of Assyr and other "limitrophic" regions in Arabia), or were instead imported into the maritime gateway to Mecca from Bombay, via the British-held port of Aden.[24]

In her seminal study of plague in the East Mediterranean, Nükhet Varlık mentions that, by the sixteenth century, sources indicate knowledge of maritime trade's involvement in the transport of rats across the sea. The stowaways were indeed considered a nuisance such that "the Ottomans had the habit of carrying weasels or cats on board of ships expressly for the purpose of 'rat control.'"[25] Still, although the concurrent absence of plague and rats was noticed, no etiological conclusion was drawn from this connection at the time. By contrast, the connection between plague and the rat was assumed by late nineteenth-century Ottoman medical experts, well before Paul-Louis Simond's scientific demonstration of the link; 1897 saw the publication of the first extensive study of plague in the Ottoman medical press since the discovery of the bacillus three years earlier. The study, authored by Dr. Nikolas Taptas, made note of the relation between the disease and rat epizootics in southern China (as related to by authors like Emile Rocher) so as to stress that "the rat is an animal that takes on spontaneously the plague more easily than all the other animals."[26] Ques-tions regarding whether plague could in fact spread from rats to humans, and, if so, in which way, remained however inconclusive. On the footsteps of similar opinions prevalent at the time across the British Empire, Tap-tas maintained that plague is principally contracted through the digestive tract, with rats being infected through eating human corpses.[27] And yet his paper also mentioned *in passim* the potential importance of the rat factor in the transportation of plague via maritime trade, a subject that resonated with contemporary suspicions of quarantine-evasion by smugglers arriving from Bombay by way of Muscat and Hadramaut.[28]

Most importantly, Taptas adopted the theory that the rat rendered supposedly soil-borne, attenuated plague bacteria virulent again, eventually leading to human infection.[29] This theory attempted to bridge two prevalent hypotheses at the time: on the one hand, that plague was borne by rats, and, on the other hand, that plague was borne by the soil. Since Yersin's discovery of the plague bacillus in 1894, the soil had been at the center of extensive debate among plague experts, with Yersin himself arguing that it should be considered a reservoir of the bacillus, which could explain recurring outbreaks of the disease in given locations: "It is possible that, in order to renew its virulence, [the bacillus] might have to make a long evolution in the earth."[30] According to this view, under certain conditions the soil functioned as the context of the development or transformation of the bacterium. In other words, the soil was considered as far more than an idle container of plague—it was its medium proper, in the sense that it was seen as giving rise to virulent forms of the pathogen after periods of dormancy or attenuation. What Taptas's theory added to this was the idea that, while they remained dormant in the soil, plague bacteria were picked up by rats and within their organisms the bacillus was able to revive and reattain its virulence, henceforth attacking humans.[31]

By the time that plague had reached Alexandria, in 1899, nobody in Istanbul seemed to doubt that it was an "Indian importation."[32] Following closely the development of the plague epidemic in British India, Ottoman doctors reproduced long-held ideas of plague as an insidious disease: "the plague virus eludes the best efforts of struggle [against it], it annuls the effects of [our] best efforts and it awaits for the most favourable moment, unknown until now to science, for emerging out of its slumber so as to assume its morbid progress."[33] Yet a key question remained: Did the disease retain or lose its force as it distanced itself from its original "soil"?

In the course of the June 30, 1899, session of the Imperial Medical Society in Istanbul, Dr. Stekoulis, the Dutch delegate to the Quarantine Council, expressed an opinion that would come to dominate Ottoman epidemiological reasoning. On the one hand, he defended the view that plague was "a disease of the soil in the same way that cholera is a disease of water," stressing the mediating role of the rat as a "multiplier of the disease."[34] On the other hand, observing the relatively "benign" character of the Alexandria outbreak, he speculated that, as the disease moved further and further from its point of origin or foyer (be that, in his mind, Hong Kong or Bombay), it lost its force and became more and more attenuated.[35]

Such theories and questions would assume particular epistemic and practical importance in light of an incident involving the boat *Polis Mytilini* in the port of Trieste in November 1899, which would destabilize the provenance of boats as an epidemiological datum and challenge prevalent notions of what constituted effective quarantine measures against plague. The boat had sailed from Istanbul and reached Trieste on October 28, after stopping at several Greek and Ottoman ports. There, a sailor reported sick with bronchial catarrh. Soon after being admitted to hospital, he developed red spots in his abdominal area and lower limbs, leading to the suspicion of typhoid fever. The patient was put in isolation where he died on November 4. Autopsy, however, seemed to confound the original diagnosis, with the presence of pyaemia leading the doctors to suspect plague instead, a suspicion confirmed by bacteriological examination.[36] What was more worrying than the rather common initial misdiagnosis, however, was the fact that the ship had not sailed from a contaminated harbor. Though seemingly an isolated incident, the *Polis Mytilini* case was catalytic in promoting calls for the deployment of maritime fumigation in the Ottoman Empire.

In the November 24, 1899 session of the Imperial Society of Medicine, Stekoulis analyzed the report sent to the Board by Dr. Stiepovich and related that the captain of the ship reported a curious fact:

> Two men having descended into the hold in order to disinfect it fell almost dead. We removed them and brought them back to life, after which we aerated the hold and wanted to remove drums containing molasses. At that moment, we perceived the presence of a large number of rat cadavers. It was without a doubt the emanations from the fermentation of the molasses in the drums, mainly carbonic acid, that provoked the asphyxiation of the rats.[37]

This bizarre incident led Pierre Apéry to comment: "We should profit from this accident for crafting a method or means for destroying rats in boats' holds. In effect, this acid unites all advantages: it is non-inflammable, inodorous, more dense than the air and does not damage goods."[38]

Apéry was the scion of one of the most powerful pharmacist families in Istanbul. Founder and editor of the two leading francophone Ottoman medical journals, *Revue Médico-Pharmaceutique* and *Gazette Médicale D'Orient*, he was also the inheritor and owner of the Grand Pharmacie in Galata, where, on account both of its products and well-stocked international library, Istanbul's medical elite flocked.[39] An investigative mind, Apéry had already noted that the *Polis Mytilini* incident appeared to be operating on the same chemical process identified sixty years earlier by Alfred Swaine Taylor as the principle behind the then very popular volcanic tourist attraction of Grotta del Cane in Foro di Pozzuoli. There, in a cruel spectacle that formed a popular part of the "Grand Tour," dogs were introduced in the Neapolitan cave (figure 3.1) so that the spectators could watch them suffocate and die from exposure to CO_2.[40] Apéry concluded that carbon dioxide could be artificially manufactured for deratization, especially onboard ships, and with the purpose of eradicating plague. In his opinion, carbon dioxide was especially fit for that purpose given the inefficiency of rat poisons and the odorous nature of other fumigating agents, which, he argued, led rats to seek refuge in their nests or outside of the hold, beyond the

Figure 3.1
The Neapolitan Grotta del Cane.
Source: Arthur Mangin, *L'air et le monde aérien* (Tours, France: Alfred Mame et fils, 1865), 162. Wikimedia Commons.

reach of the gases.[41] A few months later, in a session of the Imperial Society of Medicine on May 11, 1900, Apéry presented his thoughts on CO_2 as a means of deratization of ships in a systematic manner, attributing its inspiration directly to the *Polis Mytilini* captain's comments on the molasses incident.[42] In the context of plague having made new ravages in Arabia and in Egypt, Apéry argued, the destruction of rats on board of ships, which had for long "preoccupied the attention of commerce, navigation and sanitary science," became all the more pertinent.[43]

Apéry stressed that, by contrast to sulphuric acid, which was produced by burning sulphur directly onboard ships, carbonic acid is odorless and thus imperceptible by rats.[44] But most importantly, he noted that, if placed inside a large glass bottle where carbon dioxide was then inserted, rats died after eight minutes, with rat cadavers being preserved in a good state inside the gas for even one week, without putrefaction if the bottle was well sealed. This was considered a distinct advantage as it meant that rat cadavers would not contaminate merchandise and especially foodstuff carried in the holds of the boat before being discovered and removed. It thus allowed, first, for holds to be fumigated without goods being unloaded, and second, for a relative delay in the discovery and removal of all rat cadavers following the operation, commonly by simply throwing them into the sea. As a result, if that method were to be applied, in most cases at least, rat removal could wait until after the boat had reached its destination and goods had been unloaded onto the docks—a distinct advantage from a financial perspective and in light of the broader urge to minimize detention time.

Faced with critique by members of the Society that the removal of rats after the application of the gas would be practically impossible, Apéry responded that his method first immobilized and then killed the rat, thus not allowing it to seek refuge in the structural gaps of the boat, like other methods did. A more pertinent critique related to the production of the necessary quantity of carbon dioxide, which, in the words of Dr. Leon Fridman "would require mountains of chalk or marble."[45] Fridman explained: "for a cubic meter of carbon dioxide we need more than 2 or 3 kilos of chalk and equal amount of sulfuric acid."[46] Hence a hold containing up to 16,000 cubic meters would require more than forty-eight tons of chalk to kill the rats lying therein. Apéry retorted that in the case of *Polis Mytilini* just a few drums of fermenting molasses sufficed to kill the rats in the boat's hold.

Reminded by the presiding Dr. Spyridon Zavitziano (the US delegate to the Quarantine Council) that nobody had performed a chemical examination of the rat corpses in said boat, and hence their cause of death remained speculative, Apéry returned to the subject in the session of the Society a week later (May 18, 1900) so as "to provide satisfaction" to Fridman's observations.[47] Taking as the hypothetical volume of a boat's hold Fridman's aforementioned 16,000 cubic meters, Apéry asked how much marble it would actually take to produce the needed CO_2. He calculated that it would indeed require 68 tons, but, at a medium density of 2.7, this would represent no more than 25.5 cubic meters—only 1/640 of the overall volume of the hold in question. Hardly thus a "mountain" of marble, as Fridman had originally maintained. Moreover, Apéry argued that only a tenth of the hold had to be filled with CO_2 for fumigation to be effective, and thus the necessary and sufficient volume of marble for this operation would be drastically reduced to only 6 or 8 tons; a space occupying no more than 2.5 cubic meters. Fridman did not leave this syllogism uncontested: "If carbon dioxide accumulates entirely at the bottom of the hold, being almost in a pure state, I do not see how the air, free from carbonic acid in the beams, can kill the climbing rats there."[48] Moreover, in order for the antiputrefaction effect described by Apéry to be operative, Fridman argued, it would not suffice to simply generate enough gas to asphyxiate the rats; it would instead require a total replacement of air by carbon dioxide, something that could only be brought about by large quantities of calcium carbonate and sulphuric acid, with the help of enormous machines for the production of CO_2.

Apéry had himself considered the use of mechanized pumping of his gas into the boat holds, with the help of a Kipp generator and of a rubber tube, which would bring the gas down to the holds where it would circulate by means of properly arranging corridors between merchandise before the start of the fumigation process.[49] And yet he angrily confronted these objections, claiming that his suggested quantities "mathematically" sufficed both for asphyxiating all rats in the holds and in preserving them from putrefaction; indeed an effect that, he argued, could be brought about simply with the help of "a few enameled basins" containing "pieces of marble," and certainly not necessitating the employment of big machines.[50] Pressed by his colleagues to provide a more satisfactory description and demonstration of his method, Apéry announced that "one of the most honourable navigation companies" of Istanbul (he did not clarify which one) had expressed its interest in putting a boat at his service for applying his deratization process.[51]

In spite of this commercial interest in it, Apéry's method faced a severe limitation. This was pointed out by the member of the Belgian Royal Academy of medicine, V. F. J. Desguin, with whom Apéry maintained a heated argument: besides its deratization properties, the method had no disinfecting ones.[52] This meant that, even after deratization had been successful, any "infected cargo" still needed to be destroyed or disinfected by other chemical agents. Desguin presented Apéry with a real case, which illustrated the problems posed by this: When the steamer *Berenice* arrived from Canton into the port of Trieste, in December 1899, four reported cases of plague among its crew led it straight to the harbor's lazaretto. Composed of over thirty-five tons of coffee, the boat's cargo was worth more than two million Austrian florins. Although it had not been in contact with the crew, the cargo immediately became the object of sanitary contestation. While sanitary authorities proposed disinfection, the city's population demanded the incineration of the coffee sacs. In this dire situation, what good would Apéry's deratization method be? Desguin reasoned that "it would modify in nothing the embarrassing situation in which we find ourselves."[53]

It is thus indicative of the need to arrive at some practical solution, however incomplete, that in spite of this obvious limitation and the universally agreed importance of combined disinfection and disinfestation at the time, Apéry's method soon received broad international recognition and endorsement. The Superior Council on Public Hygiene of Belgium brushed aside Desguin's objections and his proposed alternative (deratization via carbon monoxide) so as to accept Apéry's method as "the most rational and most practical."[54] Apéry's method appealed even further afield, with the Liverpool Board of Trade declaring it the most reliable method for rat destruction in its circular "On the Influence of Rats in the Propagation of Plague" in the autumn of 1900. The Board even suggested some practical amendments to the method: placing pieces of cheese at the center of the hold so as to attract rats before introducing the gas.[55]

At the same time, and in light of a chain of new plague cases observed in Smyrna in the spring and summer of 1900, the Ottoman's Empire protection from plague assumed new importance. The fear was that Smyrna could turn into "a foyer of plague."[56] When, "finally," in September 1900,

Istanbul "had also the honour to be visited by plague," the inability to trace the origin of the infection of the patients (all of them passengers of the steamship *Niger*) troubled medical authorities.[57] Indeed the German delegate to the Quarantine Council, Dr. Andreas David Mordtmann, saw this as no less than a proof of the "complete bankruptcy" of the quarantine and cordons system.[58] These measures, he claimed, were both totally inefficient and economically disastrous: "The reform of the methods of defense against plague become day by day more urgent."[59] A similar opinion was echoed a month later when Stekoulis, celebrating the cessation of plague in what he saw as the three foyers of the disease in the Ottoman Empire (Smyrna, Beirut, and Jeddah), claimed that this was brought about by the implementation of local measures against the disease (isolation and disinfection)—the only "rational ones against an epidemic."[60] This was explicitly contrasted to quarantine, which was portrayed as an economic affliction in its own right, often graver than plague itself.

Key to the antiquarantine stance of the Society, and the subsequent call for reform of international regulations, was the oft-repeated belief that modern plague was dissimilar to plague in historical times; the main difference allegedly being that once the former distanced itself from its point of origination, it lost its virulence, assuming an attenuated or even "benign" form, as the recurrent but sporadic cases of the disease in Istanbul appeared to have confirmed.[61] However, as we have already seen, this epidemiological reasoning entailed an important contingency. The same medical authorities that maintained the nonvirulence of "distanced" plague also argued that plague was an "insidious" disease, which, on the one hand, remained largely unknown to scientists, while, on the other hand, retaining a capacity to assume its famous "frightening forms" once it had "found the necessary conditions."[62] Plague, it was argued, even when in an attenuated or benign state, hung above the Empire like Damocles' sword, ever ready to assume "its terrible side."[63] This double rationality put center stage the rat, whose role as the main carrier of the disease rendered quarantines "chimeric," urging their replacement by a "rational" system of deratization.[64] The question of quarantines was extensively discussed at the session of the Imperial Society of Medicine on November 29, 1901, under the presidency of Stchepotiew, where Mordtmann's positions on the "chimeric" nature of quarantine and the need to "bury" it, as more harmful than plague itself, were fully adopted. Stchepotiew went as far as to express the opinion that the imposition of quarantine measures against neighbouring countries are "a means of war in a time of peace."[65] Stchepotiew largely reflected a century-long tradition of mercantilist logic when he stressed that "movement is life" and that the best prophylactic against disease is the fortification of individual organisms against it.[66]

Apéry's ambition to promote his method was nested in this debate, with the editor of the *Revue Médico-Pharmaceutique* pondering "why we persist in destroying [rats] in the holds of boats by primitive processes such as sulphuric acid and other gases [which are] deleterious and more or less ineffective"; "a little less routine," he stressed, "and a little more of practical experimental spirit" and the superiority of carbonic acid "would be recognised as one of the most useful developments in naval hygiene."[67] However, we need to pay close attention to Apéry's stance toward quarantine here. As we have already noticed, his method involved deratization but not disinfection. Apéry was sceptical as regards the ability of disinfection to abolish the time-wasting and trade-hampering necessity of quarantines. This position made him very unpopular with his colleagues. Already in an editorial published in the *Revue Médico-Pharmaceutique* on November 1, 1901, Apéry openly doubted the effectiveness of disinfection, claiming that, in any case, "the germ will escape in 50 out of 100 times, often lodging where disinfectants cannot reach it" so that it might reappear when least expected.[68] The supposedly elusive, "hiding" character of plague, as well as other diseases at the time, needs to be once again underlined here. For in Apéry's ontology of the disease, this was not simply one of its traits, but instead a mechanism of nature itself aimed at "saving the race of the microbe."[69] The only solution under these circumstances would be to act as one does in any situation of war against the stealthy attack of the enemy: "to set up sentinels" who can open fire at the first sight of our foes.[70] These sentinel devices, Apéry argued, were no other than quarantine, which could push back any surprise attack. This perhaps marks the earliest instance when a proponent of fumigation openly declared disinfestation to be the sole aim of fumigation and an adequate measure against plague. Yet for Apéry this could only work in combination with quarantine as a sentinel device.

It was not long before this opinion clashed publically with prevailing ideas about quarantine in Istanbul. In the session of December 13, 1901, Apéry staged a small revolt against the antiquarantine line dominating the Society at the time. Apéry's stance on the question of quarantine was rather more nuanced than the one of Stekoulis or Mordtmann, whom he reminded that while they were ridiculing the idea of quarantine they were equally accepting the hospital or home-based isolation of the sick. In which way did the two practices differ in principle?—"Is it the word quarantine which frightens you?" Apéry asked.[71] If you think, he argued, that disinfection methods can abolish quarantine, should we also abolish the isolation of the sick? This abolition doctrine ignored the limited efficacy of disinfection, a method that was effective, in Apéry's mind, only in combination with quarantine and isolation. If his colleagues were right, Apéry reasoned, in stressing the harmful effects of quarantine, what was needed was to improve existing lazarettos, not to abolish them: "in order for commerce to profit it is not right for the merchant to perish"—for, "if movement is life," Apéry argued (referring back to Stchepotiew's antiquarantinism) "it is this movement too which is the cause of death."[72]

Apéry thus struck a middle-ground approach that saw maritime fumigation as a synagonistic rather than antagonistic partner of quarantine. Arguably this stance was miscalculated, for while his fumigation method was hailed by several quarters as successful, it ultimately failed to galvanize support from parties aspiring to an end of quarantine, and the liberation of maritime trade in the Mediterranean and the Red Sea from its politically susceptible time delays. By contrast, one of the key supporters of Apéry's method was the Ottoman court, which was anxious to maintain its system of quarantine control in spite of technological revisionism. At the international congress on maritime security held in August–September 1901 in the Belgian port of Ostand, Apéry's method was unanimously voted as the superior maritime deratization method in existence.[73] If in practice this was an endorsement of little practical effect, Apéry still capitalized on it as it attracted the attention of the inspector-general of the Ottoman Department of Sanitation and plague expert, Cozzonis Effendi, who in a letter to Apéry informed him that the Quarantine Council had decided to conduct an experiment with his deratization method with the help of the inspector of the Quarantine Council, Dr. Charles Zitterer.[74]

Soon after, Apéry received another endorsement, this time from the Sultan's chief chemist and general inspector of public hygiene, Bonkowski Pasha.[75] This lengthy and extremely flattering letter expressed the pasha's warmest congratulations for the international success of Apéry's method, and for his ingenuity in having derived the method from the *Polis Mytilini* incident. Bonkowski informed Apéry of his keen interest in the method and his belief that only CO_2 fulfilled all the desired goals of deratization without any of the usual disadvantages of other gases or methods employed on board of ships. Then, on December 30, 1901, the members of the Constantinople Board of Health as well as of various embassies and legations to the Ottoman capital received an invitation by Apéry to witness the demonstration of a rat destruction method on board of *Chios*, a Greek vessel belonging to the Aegeum Company (*L'Egée*). By the time of the experiment, Apéry had revised his previous, rather simple method of producing CO_2 by means of adding acid, such as hydrochloric acid to any carbon, such as marble, on the upper hold corridors. Instead he proposed the use of large CO_2 generators, such as the Hermann Lachapelle machine (usually used to produce steam but which Apéry thought could be retrofitted to generate CO_2 with the help of sodium bicarbonate and SO_2), either placed on piers or on smaller boats and connected to the ship's hold by means of tubes. However, Apéry maintained that the most practical solution would be to use liquid CO_2, as was the case in Marseilles. The experiment on the *Chios* appeared to be successful insofar as the rats placed in different parts of the hold were all found asphyxiated after two hours of fumigation.[76] However, as Franck Clemow, the British delegate to the Quarantine Council and plague enthusiast, noted, the experiment crucially failed to show whether the gas could reach every nook and cranny of a fully loaded hold by its own weight, leaving no place where the rats may seek refuge.

If the history of technological innovation in the late Ottoman Empire remains an underdeveloped field, the story of Apéry's fumigation method points out the way in which technoscientific debates about the relation between disinfection, disinfestation, and quarantine lay at the heart of Istanbul's emergent biopolitical apparatuses. Central to the turn toward a technoscientific management of maritime trade's public health aspects was the inclusion of animals, rats in particular, in the problematization of human health. Accounting for the swift adoption of what we would today call a zoonotic perspective of plague by Ottoman public health institutions at the end of the nineteenth century, when similar institutions in the West maintained a much more hesitant position as regards the import of human–animal interaction as regards the unfolding pandemic, requires an analysis of late Ottoman perceptions (both lay, religious, and scientific) of nonhuman animals that cannot be carried out here.[77] What is important, in terms of our study of the history of maritime fumigation, is that setting the rat and its destruction at the center of antiplague control measures allowed Ottoman authorities to develop technologies that could compete with European (and indeed North American) methods of maritime sanitation, challenging the view (both historical and contemporary) that Ottoman power simply relied on antiquated regimes of quarantine.

This was all the more important, as on the other side of Europe, in Hamburg, a fumigation technology was being developed to protect newly unified Germany against plague. Also carbon-based, but embodying distinct engineering and chemical principles, this was a maritime technology that would also place the rat at the center of its destructive attention.

Nocht's Hamburg Experiments

In the summer of 1892, Hamburg was hit by a cholera epidemic, leading to over 15,000 cases and over 8,000 deaths. The extent of the outbreak was so shocking that Robert Koch himself remarked that the sight of devastation made him "forget that I am in Europe."[78] As Richard Evans has described in great detail, the 1892 cholera epidemic marked a watershed in the public perception of bacteriology and hygiene.[79] Koch, like others before him, had shown that the cholera bacillus was water-borne. But Hamburg's miasmatists denied that the disease could have such a simple and unicausal origin. In any case, they argued that Hamburg's piped water system was new and efficient. But it was precisely the centralized but badly filtered piping system that had led to the outbreak's catastrophic dimensions. When Koch, sent to save the city, took charge, he dealt a fatal blow to both miasmatic theory and the cholera outbreak by cutting the water supply to the affected areas and by propagating strict boiling of all drinking water.

The city officials, impressed by the rapid success of Koch's intervention, reacted fast. In September 1892, a student of Koch, Georg Theodor August Gaffky, was elevated to the rank of hygienic counselor to the city and immediately set up a hygienic laboratory. Another successful student of Koch, Bernhard Nocht, at the time drafted to the German navy, proposed the installation of a permanent port physician at Hamburg's busy harbor. The city followed his suggestion and created the position, which Nocht filled himself from 1892 to 1906.

Since his nomination as port physician in the aftermath of Hamburg's cholera epidemic, Nocht aimed to establish a rigorous system of protection against the importation of diseases. Already in 1894, he complained about the lack of progressive chemical solutions, specifically designed to be applied in ships with the purpose of killing not only pathogens but all possible vectors in the hull of vessels as well as in the cargo itself. Nocht considered the application of sulphur as pointless, as he was convinced that its properties did not allow it to penetrate surfaces or densely packed goods.[80] While his early experimental procedures were heavily influenced by the threat of cholera importation, and were thus dedicated to the successful disinfection of bilge water, he would later dedicate years of work to the protection of Hamburg against plague and consequently against the importation of rats. In 1897 he published his concerns on the matter, stressing two points: a) that a simple translation of anticholera procedures would not work against plague, and b) that reinstatement of general quarantine against plague would neither provide the desired protection nor justify the immense economic cost.[81] Instead, as the recommendation from the 1897 sanitary conference in Venice stated, the focus should move toward the possible transmission of plague through rats. Yet, "the destruction of rats and mice on board," Nocht wrote in 1897, "still remains an unsolved task."[82]

At the time, vessels arriving from non-European ports were required to perform deratization upon each arrival at Hamburg, whereas vessels visiting the port on regular basis from European ports were only required to perform deratization every three months. Boats plying the River Elbe were required to perform deratization once a month.[83] The process was two-fold, comprising in combined rat-catching or poisoning, and fumigation. The former was applied to passenger and crew cabins and other small compartments, with the help of professional rat-catchers (so-called Kammerjäger). Although a respected profession, the method yielded quite mediocre results, as many rats managed to escape the hunters, whereas traps were not always effective in bringing the rats out of their hiding places in ships. Following rat-catching, the hold of the vessel was fumigated "after the receipt of the written permission of the chief harbor master, under the supervision of the harbor police."[84] Taking place directly at the place of disembarkation (and not in quarantine stations or lazarettos), fumigation proceeded after unloading the cargo, and was thus applied to empty holds. The usual process involved the burning of sulphur and charcoal in iron pots. At twenty and ten kilos respectively for every 1,000 cubic meter area, these were placed in the lower parts of the holds and burned for ten hours. An alternative method, used in cases where quicker processing was required, was the use of a newly invented chemical compound, called Pictolin.

Invented by the Berlin-based Swiss physicist, Raoul Pictet, Pictolin consisted of sulphur dioxide and 3 to 4 percent carbon dioxide and was liquefied under pressure and then delivered in iron bombs. Comparable to the Clayton machine, it was introduced into the holds of vessels via a system of generators and tubes. With twenty kilograms of Pictolin, about 1,000 cubic meters of space could be freed of rats. It was originally invented to further Pictet's development of fridges and refrigeration technologies. It was its surprisingly efficient capacity in killing rats—probably due to its high ratio of sulphur dioxide—that led it to be included in German disinfection efforts. Its unique stench promised an additional layer of safety for humans, as its presence was (unlike that of carbon dioxide) clearly recognizable. Once a Pictolin "bomb" was opened, it operated by quick evaporation, leading to the replacement of air with its asphyxiating gas. The gas was usable only on empty holds, with noninflammable properties, a trait that was a great advantage in spite of the greater cost of the process.[85]

Nocht had amassed a prestigious amount of experience in India, where he had worked with Robert Koch, and developed a keen interest in the emerging discipline of tropical medicine.[86] Chief Medical Officer of the Harbour of Hamburg since 1893, he became eventually the director of the newly founded *Institut fuer Schiffs-und-Tropenkrankheiten* (Institute for Maritime and Tropical Diseases, from October 1900). As his field of expertise enjoyed institutionalization, he was keen to also set methods of fumigation on sound scientific grounds. Instead of trusting the already established methods of deratization, in 1899 he began to devise a more sophisticated system to destroy rats on ships, using a mixture of carbon monoxide and carbon dioxide.

A fundamental problem with Pictolin at the time was not only the enormous cost involved, but also its inconvenient nature as regards a long list of valuable merchandise. To fumigate with Pictolin, Nocht remarked, required the merchandise to be fully unloaded, which posed a risk of escaping rats, and required additional methods to disinfect the merchandise by other means.[87] Instead, Nocht developed and refined his own method of utilizing

CO_2. In close collaboration with Robert Koch, who was interested in ascertaining the best fumigation method, Nocht undertook various experiments from 1899 onward, employing gas produced by the reaction of hydrochloric acid on marble. The looming question for Nocht concerned the correct density of the gas for arriving at the desired result: disinfestation. Although the Hamburg experiments with CO_2 were conducted even before reports of the *Polis Mytilini* reached the German port city, Nocht remained critical of Apéry's method of testing the asphyxiating quality of the compound. To sink little candles into the holds of fumigated ships and to judge upon their expiration whether the amount of oxygen was sufficiently low, seemed to Nocht inconclusive when testing the gas's capacity to kill rats. Apéry's reports, Nocht complained, did not provide detailed or systematic descriptions of experiments, which could be repeated and tested with similar configurations in different places. Furthermore, Nocht considered the use of carbon dioxide as problematic, as its gaseous distribution was difficult to maintain and progressed only very slowly.

Instead Nocht aimed to bring fumigation into more sustainable application with a special mixture of CO and CO_2: Kohlenoxydgas. The gas had no smell, it did not damage the merchandise, yet it was poisonous enough to be lethal even in small dosages. Furthermore, it was distributed rapidly and led to a quick onset of paralysis in rodents before their imminent death. Nocht initially tested the introduction of 20 percent CO_2 into a loaded hold, where cargo was estimated to be occupying 50 percent of the total hold volume. A few years later, the 1903 report of the International Sanitary Conference in Paris detailed his experimental setup:

> He had three hundred rats enclosed in cages in the empty holds of a large boat, *Bulgaria*. These cages were distributed on different points and were entirely covered with mattresses, bags and similar objects piled up on a large height. Other cages were placed in holes. The generating gas containing carbon monoxide was then sent and all the rats were killed. This experiment was renewed with liquid sulphurous acid, but then the rats remained alive.[88]

In the years to follow, Nocht refined and improved his system of CO_2 fumigation, and in 1903 he published a lengthy report, which demonstrated the superiority of his apparatus in comparison to the Clayton machine but also to other CO_2-based methods tested by Haldane in the UK.

Nocht's fumigation machine was built in collaboration with the Berlin-based J. Pintsch Company and was installed on a floating platform

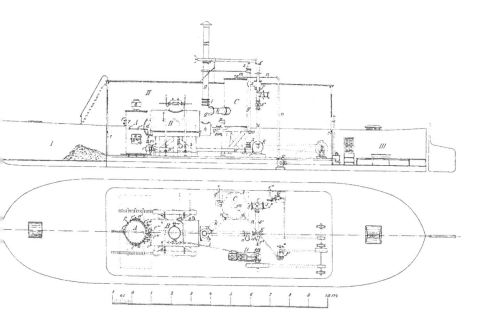

Figure 3.2
Plan of Nocht's fumigation machine.
Source: Bernhard Nocht and G.Giemsa, "Über die Vernichtung von Ratten an Bord von Schiffen. Als Massregel gegen die Einschleppung der Pest," *Arbeiten aus dem Kaiserlichen Gesundheitsamte* XX, no. 1 (1904): 98.

(figure 3.2). The aim was to provide enough gas of the highest quality, to succeed in the total eradication of rodents, but to also provide a mixture that would prevent the gas from becoming an explosive compound ("Kohlenoxydknallgas"). As it was produced in a simple generating furnace, Nocht called his invention "Generatorgas."

The gas was produced by burning charcoal through the injection of large amounts of oxygen. Part of the heat was used to operate a steam engine to drive a water pump and a ventilator. The apparatus was able to inject gas, as well as to suck the remaining gas out of the holds of vessels. After the gas was generated in the furnace, it was cooled and cleaned through steam and sprinkling water. The gas was designed to have at least twice the amount of CO_2 compared to CO and was thus considered to be noninflammable.[89] The average proportions were 4.95 percent CO, 18 percent CO_2, and 77.05 percent N. The introduction of the gas into vessels was carried out in comparable manner to the existing practices of the Clayton machine: The apparatus would be attached to the existing vessel-borne tubes, while all openings were roughly closed with fabrics and cushions (figure 3.3). The only substantial difference was to be found in the imminent and lasting danger of the undetectable gas, which was, even at a 0.5 percent density, capable of leading to substantial poisoning and possible death in humans. Recommendations were therefore given for an extensive circulation of fresh air and for testing the safety of the holds through mice, lowered down to the holds inside cages.[90] Detailed experiments with infected rats in ships like the *Bulgaria* would prove that the apparatus was capable of destroying all rats and mice in every part of the vessel, even in hiding places under large quantities of goods of all kinds. Being absolutely nonhazardous to goods, holds, and the vessel's structure, the Generatorgas had proved, so Nocht claimed, to be the best instrument against the introduction of plague.[91]

However, a persistent problem with the machine was its tendency to leak gas at various points within the piping system. Due to the lack of smell and visibility, the leakages led to a series of dangerous incidences including three fatalities. This prompted Nocht to redesign the apparatus, allowing for the gas to be less pressurized inside the machine, owing to an improved arrangement of the exhauster. Passed Assistant Surgeon Victor G. Heiser from the U.S. Public Health and Marine-Hospital Service observed the apparatus in 1909 and reported his favorable impressions. In particular, the capacity of the machine to produce over 3,000 cubic meters of gas per hour was noteworthy, especially in combination with the low cost of material required to generate the gas from coal. Heiser particularly emphasized the utility of the apparatus for all cargo that was shown to react unfavorably to fumigation with sulphur, including camphor, silk, and tea, as these appeared to be left unharmed by the Generatorgas. Most importantly, the machine seemed to be fulfilling its prophylactic purpose: being used in the fumigation of twenty-one plague-infected vessels and leading to the death of 171 verified infected rats, it was praised as the reason why not a single human plague case had been reported in Hamburg.[92]

Ottoman and German carbon-based fumigation technologies consolidated the rat as the prime target of maritime fumigation well before the particular animal had been scientifically stabilized as a host of plague. Yet lacking a mechanically sophisticated application, Apéry's machine never competed with other fumigation apparatuses on the global stage.

Figure 3.3
Photograph of Nocht's fumigation process applied to a vessel in the port of Hamburg.
Source: Bernhard Nocht and G. Giemsa, "Über die Vernichtung von Ratten an Bord
von Schiffen. Als Massregel gegen die Einschleppung der Pest," *Arbeiten aus dem Kaiserlichen Gesundheitsamte* XX, no. 1 (1904): 106.

By contrast, Nocht's machine was to become the main rival of the Clayton machine. In this way these two carbon-based fumigation technologies
brought together quarantine, disinfestation, and disinfection in ways that
catalyzed technoscientific visions of maritime sanitation that would dominate European visions of hygienic modernity in the opening decades of the
twentieth century.

4 Clayton and the "Defense of Europe"

With his patent granted in September 26, 1899, Clayton moved his business to New York City, where he incorporated the "Clayton Fire Extinguishing & Disinfecting Company of New York." His machine was immediately praised in *Engineering Magazine* for the ingenious ways in which temperature and oxygen supply could be controlled while the machine was running, allowing for rapid reactions and careful adjustments to a variety of scenarios.[1] Clayton's company became swiftly the sole supplier of fumigation machines to the Federal Marine Hospital Service, and his machines were introduced into the ports of Philadelphia, New York, and San Francisco.[2] From here, the Clayton machine emerged as a standard method of fumigation against plague, yellow fever, and other infectious diseases carried across the sea. Over the following years, he registered a holding company in West Virginia as "Sulphur Dioxide Fumigation and Fire Extinguishing Company" to manage his global endeavors. The company was incorporated with a capital stock of $1,000,000, which was doubled to $2,000,000 in 1901. Indicative for the astounding success of his overseas endeavors, Clayton established a second company in West Virginia, which was incorporated with an impressive capital stock of $5,000,000 in April 1900. Among the registered directors ranked the former president of the Louisiana State Board of Health, Samuel R. Olliphant. Clayton himself remained acting President of his growing global business until 1909. After 1900, he began to open a series of international branches. He founded the "Compagnie du Gaz Clayton Company of France" in Paris and the "Clayton Gas Company" of Egypt. But his most successful overseas endeavor would eventually become the "Clayton Fire Extinguishing and Ventilating Company" of London.[3]

As Krista Maglen has noted in her history of quarantine and immigration in England, "by the late nineteenth century, Britain's ports were teeming with ships from around the world. The empire's prosperity and power were at their peak, and commercial and passenger shipping reached levels never before seen."[4] Maglen calculates that by 1892 over 10,000 ships were arriving in London alone on an annual basis from overseas harbors. Through the innovative introduction of the Port Sanitary Authorities in 1872, British ports relied increasingly on "a combination of port sanitation and sanitary surveillance."[5] This so-called English System was intended to meet the needs of the increasing traffic in maritime trade, by basing quarantine and isolation not on the existence of disease at the port of origin, but rather on its direct observation by port authorities at the moment of arrival. Eventually, in 1896, the Quarantine Act was repealed in Britain, including for diseases like plague and yellow fever, for which the United Kingdom's legislation had specifically been designed. However, Maglen notes, as quarantine continued to prevail in the control of the international traffic of both humans and merchandise across the globe, "the English System remained in the shadow of quarantine legally and internationally."[6]

Whereas, as Graham Mooney has shown, disinfection formed an important part of British sanitary intervention on *terra firma*, maritime fumigation appears to have been less well established as regards holds and their cargo, than on the deck of vessels. According to Maglev's calculations, of 362,823 vessels inspected in the Port of London between 1873 and 1893 only 451 were fumigated.[7]

Introducing the Clayton to Britain

The first notification of Downing Street on the disinfecting properties of the Clayton apparatus appears to have been made on April 29, 1901, when the British Colonial Office received the copy of a letter written by the illustrious "father" of tropical medicine: the Scottish physician and founder of the London School of Hygiene and Tropical Medicine, Patrick Manson. The letter enclosed a report by the Superintendent of the British India Steam Navigation Company, Captain George Hodgkinson, on experiments carried out with the apparatus in London. Urging further enquiry into this "practicable and promising method," Manson offered his unambiguous endorsement, stating that the experiments "seem to show that by this apparatus ships can be effectually cleared of rats and similar vermin without damage to cargo and at small cost."[8]

Hodgkinson's presentation of the Clayton machine had all the characteristics of an introduction of a machine that held the promise of revolutionizing maritime trade. It stressed that the fire-extinguishing and ship-disinfecting properties of the apparatus were based on a common principle: "The rapid generation and application of sulphur-dioxide gas (SO_2) which being heavier than the atmosphere will quickly replace the natural atmosphere of a hold being assisted by the action of the return pipe which sucks the air from the hold to the generator and converts it into SO_2."[9]

Hodgkinson explained that on March 26 and April 3, 1901, a demonstration of the two properties of the machine had been undertaken on the Orchard Wharf and on *S.S. Manora* respectively.[10] The report focused on the apparatus's disinfecting properties, stating that "Each hold and each section of the passenger accommodation was filled with the gas with the result that great quantities of cockroaches and 301 rats were destroyed. It is important to note that all animal life on finding itself choking with the gas deserts its lurking place and comes out into the open to die and the remains are therefore easily removed."[11]

Even more promising was the apparently harmless nature of the process as regards cargo. Samples of coffee, tea, sugar, flour, cocoa, salt, vegetables, a polished velvet upholster chair, broads painted with zinc, and other material were tested, with only cocoa and flour reported as "slightly affected."[12] Finally, on April 19, the saloon accommodation of the boat was treated with "samples left for 24 hours exposed to the air and then tested"; the only tarnish was on the saloon's gilding.[13]

The results, Hodgkinson concluded, proved to be "highly satisfactory," as "by the use of one of these machines ships could be kept almost, if not entirely free of vermin."[14] A few days later, a second experiment, this time at the London Docks, replicated the results. Hodgkinson reasoned that, for a time when the disinfection of vessels arriving from plague-infected ports was "of such vital importance to eastern shipping interests," the Clayton process promised a fast and efficient solution to the mutual benefit of traders, shipowners, passengers, and governments alike.[15]

Both the lay and scientific press were quick to cover the story, with *The British Medical Journal* concluding that the opinion formed in light of the spring experiments was so favorable that "the system is likely to be introduced into Indian ports."[16] At a time when not only trade but also plague, and their entanglement, were recognized as being irreducibly international and maritime-dependent, newspapers like Edinburgh's flagship daily, *The Scotsman*, and the shipping magazine *Fairplay* hailed a major breakthrough.[17]

Perceptions of the silver-bullet qualities of the Clayton machine were fueled by a robust if aggressive promotion campaign by the Clayton Company, which had established its offices as "The Clayton Fire Extinguishing and Ventilating Company," at 22 Craven Street, London, in the beginning of 1901. The Company provided British authorities with comprehensive booklets, containing information on the functioning of the machine, and extracts of scientific articles, endorsing its efficacy (including such as originally published in Louisiana with regard to the Olliphant machine) (figure 4.1). It also maintained a sustained correspondence with them on various aspects of the apparatus.

Figure 4.1
Two types of the Clayton machine from promotional pamphlet, "The Clayton Fire Extinguishing and Ventilating Company, Limited."
Credit: The National Archives (UK), ref. MH19/274.

In a lengthy letter dated October 29, 1901, the Company developed an iconic argument insofar as it exercised pressure on the British government from a range of perspectives.[18] Noting the urgency of the adoption of plague-protection measures and underlining the relation of the disease to rats, the Company argued that the measures proposed by the Port Sanitary Authority of London at the time, consisting in rat-trapping and poisoning, were grossly inadequate. Even less effective, it was argued, was the alternate method advised by the Authority: burning sulphur in the vessel's holds. Not only was this method unscientific, as its efficacy was limited to the combustion allowed by the existing oxygen in the hold, but it also risked causing a fire, which could consume the entire cargo. By contrast, the Clayton machine was presented as a revolutionary invention, as it operated on the basis of scientific engineering principles: first, in that it was pressure-driven, and second, in that rather than merely charging the hold's atmosphere with gas, the machine *converted* it, by means of a return pipe, into SO_2.

The *Sénégal* Incident

The reach of the Clayton machine was not, however, limited to Britain. French concerns about rats as propagators of plague had been growing in scientific investigations of plague ever since Alexander Yersin's initial work on the disease's pathogen in Hong Kong in 1894. The Pasteur Institute was at the forefront of research regarding rats in the context of outbreaks in British India, which in 1898 led to the articulation of Paul-Louis Simond's theory regarding the role of the rat flea in plague transmission.[19] What, however, functioned as a catalyst of French interest in rats, and their relation to maritime trade in particular, was an incident not in Asia or even in the North African territories of the French Empire, but instead in continental France itself.

The incident that "underlined anew the question of plague prophylaxis" involved the *S.S. Sénégal*.[20] The ship, which had been previously used to transport spectators to the first Olympic Games in Athens, had been in the course of a leisurely cruise organized by the leading French scientific journal *Revue générale des sciences pures et appliquées*, carrying 174 passengers, including seventeen doctors.[21] Then, on September 16, 1901, two days after leaving Marseilles, and with Lipari in view, a plague case was discovered in a member of the boat's crew named Marius Fabre.[22] Within two days the boat was back to Marseilles, where its illustrious passengers, including the future President of the Republic, Raymond Poincaré, were kept in quarantine in the Frioul lazaretto.[23] Having previously arrived in the noncontaminated port of Marseilles (August 28, 1901) from Beirut with a stop at the contaminated port of Alexandria in Egypt, the boat had been considered free of plague on the basis that no human cases of the disease had been observed on board.[24]

The scandalous event made headlines and galvanized medical opinion regarding the question of the role of the rat in the propagation of plague.[25] Did rats indeed pose a danger to the health and wellbeing of Europe? Was it true that they could not only maintain plague onboard ships (with the presence of the disease being overlooked in the temporary absence of human victims) but also mix with native rats and contaminate them after disembarking from vessels? Could they thus create long-term reservoirs of plague in French and other European harbors? Or were they, as sceptics declaimed, simply victims of blanket accusations—scapegoats held responsible for the "crimes" of the true, but less loathsome, plague-spreaders like grain, soil, or clothes?[26] Entwined with these scientific questions was an array of administrative ones. Given the growing consensus that rats did play a role in the transmission of plague, and as sulphurization of vessels had been part of government maritime regulations since July 1899, why had no deratization measures been followed in Marseilles?[27] If plague on the *Sénégal* was only one among several similar incidents in preceding months, was this to be attributed to the "apathy of navigation companies"?[28] Or, as Jules Bucquoy, the doctor sent to attend the plague-stricken boat, seemed to suggest, should it be blamed on the inadequacy of the existing regulations regarding deratization?[29]

Supporting Bucquoy's thesis, Henri Monod, director of public health at the Interior Ministry of the Republic, proposed that the rules applied to cholera did not offer sufficient protection against plague, and that the lesson of the *Sénégal* suggested that the sulphurization with the purpose of rat destruction should be made obligatory for all vessels arriving at French ports.[30] On September 26, 1901, Monod issued a ministerial circular, which drew sharp lessons from the recent crisis. The circular stressed that, while existing maritime regulations regarding deratization were an important precedent, a renewed focus on the *total destruction* of rats on board of all vessels derived from infected ports, even those on "non-infected boats," by means of sulphurization was urgently necessary.[31] Monod stressed the need for this procedure to be followed closely, and to avoid reloading the boat before disinfection was complete. As a result, the sulphurization of boats deriving from plague-infected ports, after the unloading of their cargo, was made obligatory.

What was meant by sulphurization in this context is not made explicit in Monod's circular, but becomes obvious in the inclusion of a report on a deratization process employed a few months earlier in the Egyptian port of Alexandria, which Monod directly praised as exemplary. The report, written by the director of the Alexandria Quarantine Office for the president of the Quarantine Council of Egypt, stated that, following the unloading of goods from the boat's hold, sulphur was prepared for burning: after creating a thick bed of earth carrying a 20–40 cm^3 piece of charcoal, the latter was set aflame with the help of petroleum or tar, upon which sulphur was added at a quantity of 8–10 kilos.[32] The burning sulphur hence produced asphyxiating gases within the hermetically closed hold and was thus maintained in locus for twenty-four hours. Then the hold was reopened, and after one hour had passed crewmembers descended to pick the dead rats. These were in turn incinerated with the help of cattle manure in the nearby lazaretto. Following this deratization process, the port authorities proceeded to disinfect the hold by means of quicklime applied to its floor, and "phenic acid" (carbolic acid also known as phenol) on the ceiling and walls of the hold.[33]

It was this combined method of disinfestation and disinfection that the Marseilles port authorities reported as having been applied, in compliance with the September 26 instructions, to seventy-one boats out of a total of 132 disembarking in the port between September 26 and December 21, 1901. The authorities stated that, as all other boats were carrying goods in transit, sulphurization was not applied to them and they were allowed to sail without being fumigated.[34] For besides damaging metallic surfaces, this method had the major disadvantage of being suspect for damaging goods. Hence, even in the cases where it was undertaken, goods were first unloaded on the dock, with the danger being that rats removed together with cargo would then seek refuge in buildings or sewers nearby.[35]

As a result, in order to both avert this danger, and to minimise the duration of the fumigation procedure, and thus placate shipowners, the

application of liquid carbonic acid to loaded holds was tried in Marseilles
by Dr. Catelan under the supervision of Dr. Jacques:

> On the deck is placed a barrel containing hot caution at 50 degrees in which five
> bottles of liquid carbonic acid are placed at once. The hot water serves to prevent
> freezing of the gas as it exits the bottle. Each bottle contains 10 kilograms of lique-
> fied CO_2, or 5 cubic meters of gaseous carbonic acid. To these bottles, screw the
> coupling nut (not universal screw) of the rubber pipes 2 centimeters in diameter
> and about 6 meters long, intended to conduct the gas in the hold. The pipes
> plunge into the hold at a depth of 1 meter and pass through the flange of a panel.
> A bottle empties in less than three minutes.[36]

Employed in this manner (including some alternations), five fumigation
experiments were undertaken between December 4, 1901, and March 25,
1902. The aims of these ranged from seeing whether CO_2 could, in spite
of its density, penetrate the entire hold at a concentration of 25 percent,
to ascertaining the time needed to kill rats, and whether the introduction
of gas at different intervals affected the ability of CO_2 to spread homoge-
neously across the hold. In spite of this extensive experimentation, however, the trials remained inconclusive on account of conflicting opinions
regarding the exact concentration of the gas needed to effectively deratize
a boat on empty and full holds.[37]

Still haunted by the *Sénégal* incident, a session of the National Academy
of Medicine was held on March 18, 1902, to discuss the sanitary services in
the Frioul lazaretto. The discussion quickly developed into one regarding
deratization, with Charles Louis Alphonse Laveran (the discoverer of the
protozoan cause of malaria) endorsing sulphurization by means of sulphuric acid as superior to CO_2. However Laveran's position was countered by
the influential founder of the *Revue d'hygiène et de police sanitaire* and author
of the 1882 *Traité des désinfectants et de la désinfection*, Émile-Arthur Vallin,
who stressed that the misapplication of such fumigation measures in even
just one case would cost several hundred thousand francs.[38] Who would
then take responsibility for such disaster, including not simply financial
cost but also the fact that it would discredit the whole fumigation process?
Moreover, issuing such orders was a mute verdict, Vallin reasoned, when it
was not even clear if the correct amount employed in this fumigation process should be forty grams (as used in Marseilles) or ten grams (as employed
in Hamburg).[39] Being an old but disappointed partisan of sulphuric acid,
Vallin declared himself unable to share Laveran's optimism. Yet not all was

bleak in Vallin's opinion, as from the United States a new method beaconed as a hope whereby such primitive fumigation methods could be overcome—a method that, he urged, needed to be tested by the French sanitary authorities as a matter of great urgency: the Clayton machine.

The Dunkirk Experiments

Resulting from the aforementioned meeting of the Academy, on April 12, 1902, a ministerial circular boldly decreed that the obligatory sulphurization of boats arriving from plague-infected ports should be performed on loaded holds. This led to further calls of making a specific method of fumigation under the control of public health authorities legally binding for navigation companies.[40] It was, however, another minor incident that in fact triggered the first application of the Clayton method in France.

Arriving in Dunkirk on June 10, 1902, from Calcutta (via Aden, the Suez, and Malta) the British vessel *City of Perth* reported two plague patients among its crew, both of whom died the following morning. As a result, French authorities ordered the boat out of the port, forbidding any communication with the shore until, on June 17, it was decided to sail to London, but it was ordered by British authorities to anchor in the lower part of the Gravesend Reach. As all indications pointed at infected rats being the source of the outbreak, there, the boat's holds and cargo were subjected to fumigation by means of the Clayton machine. Dr. Adrien Loir was allowed to follow the procedures as a member of the French health authorities.[41] More than two hundred rat carcasses were consequently removed with the help of tongs and burned in the boat's furnace after being immersed in corrosive sublimate. Following the combined method of disinfestation and disinfection, the holds were then also "washed down by a powerful hose with disinfecting fluid."[42]

This was a most embarrassing situation, for, in spite of all the aforementioned circulars and legislation, a vital French port like Dunkirk was internationally exposed as being unable to provide fumigation services for infected boats. Saint-Nazaire, the closest lazaretto, was a three-day journey away, but it too did not possess appropriate fumigation facilities.[43] This "painful incident," as it was plainly described by French medical authorities, led Dr. Duriau, the public health officer in charge of the incident in Dunkirk, to assert: "this [the Clayton] is the machine that we need to address ourselves to defend our maritime frontiers against exotic epidemics."[44] Petitioning the Inspector of Sanitary Services of the Ministry of Interior, Paul Faivre, permission was thus granted for Clayton apparatus trials to be undertaken in the port of Dunkirk.

This was not strictly speaking the first trial of the Clayton on French territory. A few months earlier, experiments on maritime rat destruction had been led by Drs. Langlois and Loir of the Pasteur Institute's branch in Tunis. Their endorsement of the Clayton, however lukewarm, caused Apéry's acute reaction in Istanbul, when he demanded that the International Board of Health "rejects [sic] all systems based on sulphuric acid, whether these be pot systems or the Clayton system, as rat destruction methods that are illusionary and expensive."[45] During the experiments in Tunis, use of CO_2 was compared to Clayton-generated SO_2. It was concluded that the general suspicion against the latter, on account of Robert Koch having demonstrated that SO_2 could not kill anthrax spores, was misguided when applied to other pathogens. Hence, Langlois and Loir maintained that, with the exception of spore-bearing microbes, sulphurous acid gas was indeed useful against diseases like cholera but particularly against plague, as it was so much better at killing rats and their fleas.[46] The report was revealing on many fronts. For example, it showed that Pictolin was too weak for effective deratization. More importantly perhaps, it also reproduced an account by the eminent Alsatian chemist Daniel-Auguste Rosenstiehl regarding the chemical properties of the gas generated by the Clayton machine, which, for its French audience, produced a certain epistemic stabilization of this fumigant.[47]

However, back in Dunkirk, not content with their colleague's dictum, "Let us stick to sulphurous acid gas until we find a better gas to replace it," Duriau and Albert Calmette (then director of the Institute's branch at Lille) decided to undertake a series of trials with the apparatus, seeking to test so-called Claytonization on a range of microbial agents.[48] This was indeed the first time that the germicidal properties of the machine would be put to the test in Europe. The vessel of choice was the 1,200-ton iron steamer *René*, moored in the port of Dunkirk and carrying barley from the Algerian port of Oran. The experiment took place on September 27, 1902, with the Clayton machine being placed on a barge alongside the steamer.

With the collaboration of the Chief Chemist of the Ministry of Finance, the Pasteur Institute in Lille prepared typhoid fever, cholera, and plague samples with which Calmette saturated strips of flannel.[49] This artificially infected material was then placed in both dry and moist condition in cylindrical tubes (30mm diameter) which had been previously sterilized and were kept open at both ends with cotton stops attached. At the same time, other similarly impregnated strips (both moist and dry) were "packaged" in sterilized blotted paper, in sterilized flannel, and in doubled oiled foolscap, which provided the researchers with unique packages per microbial agent. Parallel to these preparations, moist and dry controls were kept on the deck, unaffected by the experimental vapors.

Two experiments were performed with these preparations. The first took place in the after hold of the boat where the microbial samples were placed in both the lower hold and in the upper part of the 'tween deck hatch. After closing the upper deck hatches, with the help of two pipes (for suction and discharge) the Clayton machine was led to introduce its vapors for two hours and fifteen minutes at a concentration varying between 8 and 15 percent. Opening the hold after two further hours, scores of rats were seen scurrying in the 'tween decks. In his report to the Ministry, Duriau explained: "As soon as it is possible to go down into the slipway, 15 dead rats were collected: Twelve other corpses of rats, plus that of a cat which can not be traced back, were discovered the following day. The floor of the cabin was strewn with dead flies."[50]

No harm was observed on the structural elements of the boat or on the furniture, including leather armchairs, whereas various types of merchandise that had been placed in the boat's hold in order to test whether the Clayton gas had damaging properties were tested and found to be unharmed. This included, among other things, beef, cheese, carrots, potatoes, and tobacco. Only exposed water and wine were judged considerably altered.

The second experiment was conducted in the deck cabin, a 250 cubic feet space containing two fully fitted bunk beds where gas was similarly inserted for two hours and forty-eight minutes. In this case, the concentration of the gas was found to be unequally distributed with SO_2 easily escaping the room, as this had not been hermetically sealed. Samples from both experiments were processed in Lille the following day by developing cultures at 37°C. In seven days, from among the experimental sample, only the dry typhoid bacilli from the 'tween deck packages demonstrated a development of microbes; a fact attributed to a possible fault in the sealing of the upper hatch, near which the said samples were placed. By contrast, all control samples with the exception of cholera in a dry state developed cultures. Calmette enthusiastically concluded that the success of the gas with regards to all microbes tested pointed to the necessity to make the use of the Clayton obligatory on all vessels "in the shortest possible time."[51]

The success of the Clayton, in Calmette's eyes at least, did not, however, mean that other modes of fumigation should no longer be tried and experimented upon, contending for the primacy of French deratization methods and official endorsement. For example, Dr Sené, director of health at Pauillac, reported that on July 17, 1902, the deratization of the cargo ship *Matapan* was undertaken in Bordeaux. The boat carried a range of products including sacs of corn, cocoa, tobacco, and barrelled sardines. After removing tobacco and cochineal bags, at the request of the merchant company, 400 kilos of sulphur were placed in furnaces allocated on different compartments of the hold (figure 4.2) and were left to burn for twenty-four hours approximately. With no impact on the cargo goods, forty-four rats were found asphyxiated and forty living, out of which only one adult animal. All living rats were found under sacs, which were believed to function as barriers to the fumes. This led Sené to suggest that, in the case of infected boats, the process should be extended up to seventy-two hours to make sure all rats have been asphyxiated.

As a result of the different and indeed conflicting opinions voiced by scientific experts and port authorities across France, Adrien Proust and Paul Faivre were charged with composing a report on different deratization methods. Presented on November 15, 1902, the report compared and contrasted three methods: sulphurization by burning sulphur in free air, the Clayton process, and CO_2-based fumigation.

In relation to the Clayton, Proust and Faivre sought to secure more information by Dr. Souchon, the President of the Louisiana Board of Health at New Orleans. He replied on March 14, 1902, confirming that the Clayton machine had been in operation in Louisiana in the last ten years, its "chief objective [being] the destruction of germs; secondarily to obtain the destruction of rats, mosquitoes and vermin."[52] Asked regarding the duration of the procedure, Souchon replied that the production of gas took between one and six hours, according to the vessel's dimensions, where it was thereafter retained for twelve to twenty-four hours. As in the case of Louisiana, the apparatus was aimed principally at yellow fever, and the

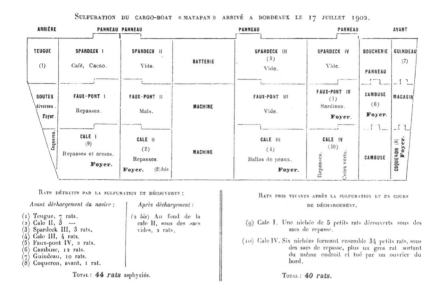

Figure 4.2
Diagram of the fumigation of the *Matapan*.
Source: A. Proust and Paul Faivre, "Rapport sur different procédés de destruction des rats à bord des navires," *Recueil des Travaux du Comité Consultatif d'Hygiène Publique de France et des Actes Officiels de l'Administration Sanitaire* 33 (1903): 341. gallica.bnf.fr / Bibliothèque nationale de France.

boat was kept in quarantine for five days, the disease's incubation period. As for the cargo, this was confirmed by Souchon to remain in the holds during fumigation, adding that "ships trading regularly with New Orleans have in each hold a special pipe to facilitate the introduction of the gas."[53] Asked whether merchandise have been observed to sustain any damage, Souchon replied that "no damage has been caused to the merchandise in a dry state. In the presence of humidity the anhydrous gas is converted into H_2SO_3 which has bleaching properties."[54] Souchon detailed that his station possessed two Clayton machines, one of which was operated by the side of the target boat with the help of a barge, while the other was fitted to a rail track running along the quay. Finally, asked about the price of the operation, Souchon replied that "The cost of complete disinfection of ships is 105 dollars (around 530 francs)."[55]

Comparing the three methods, the two doctors thus found sulphurization by burning sulphur in open air to be effective in killing rats: a simple, but inflammable method. But, as the process required unloading, so as not to harm the cargo, many rats were able to escape before fumigation even began. At the same time, unloading and reloading cargo was considered to be heavily reliant on labor. For example, in the case of the sulphurization of the Couldson in the Havre (June 28, 1902), the unloading and reloading of 2,000 sacs required the manpower of twenty-two. This had the added disadvantage that workers, generally averse to working in quarantine stations, required higher payment, while the crew had to sleep the night outside the boat; an arrangement described by its captain as most disagreeable. Finally, this gas was judged as having no penetrating properties in lack of pressure, so even if merchandise was present and left undamaged, the gas would fail to affect any but the top parts of the load. Thus the method of sulphurization by burning sulphur in open air was held to be unreliable as regards the total destruction of rats on a loaded hold. In these terms, Claytonization was considered to be a superior method, both for its ability to operate on a loaded hold, the noninflamability of the gas produced, as well as the rapidness and greater penetration force of the operation. When it came to comparing the Clayton with CO_2 methods, Proust and Faivre noted the importance of SO_2 being perceptible in terms of its odor by the vessel's crew. The example given was a man who had climbed on the vessel *La Marguerite* in a drunken state and slept there without being perceived by the Dunkirk harbor crew, who proceeded to fumigate the steamer. The man, who awoke from the smell of the gas, would have surely died had the boat been fumigated with odorless CO_2. Another major disadvantage of CO_2-based fumigation was judged to be the fact that, according to Drs. Catelan and Jacques' own confession, this could only be used in the hold proper, with application to other compartments of vessels necessitating "to modify the present mode of construction of ships."[56] However, what made the Clayton clearly superior was its ability to kill both microbes and so-called vermin, like fleas, which, as shown by Langlois and Loir's study, CO_2 was unable to do. It was only with SO_2 that the "double action" of disinfestation and disinfection was achievable. Proust and Faivre thus concluded: "The use of the Clayton apparatus, while suppressing these manipulations, brings into contact with the whole cargo of the ship a gas possessing more powerful bacteriological properties than those of other gases used up to that time. We are therefore logically led to give it in all respects the preference."[57]

Although, on the basis of this recommendation, on June 20, 1903, a circular would eventually announce to the chambers of commerce that the use of the Clayton machine under public health authority supervision had become obligatory, the two doctors were not oblivious to the limitations of the method. Indeed they had direct experience of the latter during operations undertaken on two additional empty and seven more loaded boats in Dunkirk. Though no cargo, including foodstuff, was reported as damaged by the process, the results were mixed insofar as in several cases surviving rats were found in the holds. This led Proust and Faivre to note that the results "undoubtedly leave much to be desired; however, they do not seem to us to invalidate the method, but the insufficient application which has been made of it."[58] The problem, in the opinion of the authors, seemed to be the duration of the fumigation process, which they estimated to be on average seven hours, inclusive of time taken for warming up the machine.

Responding to the report, Drs. Duriau and David hastened to remind the French medical community that the principal aims of fumigation should take into account the mutual interest of the public and seafaring capital. A "disinfecting product" should hence combine the following traits, all of which were met only by Claytonization:

> From the point of view of public hygiene [*salubrité*] it should destroy microbes, rodents (rats, mice, etc.), insect and in a complete manner vermin;
> From the point of view of navigation companies, it should not damage the ship, its furnishings, and should not necessitate the disembarking of the crew;
> From the point of view of the shipowner, it should not alter the merchandise;
> From the point of view of the insurance company, it should not cause fire;
> From a general point of view, it should be able to be employed without even partial unloading of cargo, to not delay but for a few hours commercial transactions, and, finally, to not be too expensive.[59]

These conclusions, and the optimism they carried with them regarding an effective application of fumigation in the control of germs and disease vectors, came in the aftermath of a major European crisis regarding deratization and its impact on maritime trade.

The Ottoman Deratization Crisis

The crisis concerned an Ottoman circular (No. 180, "Instructions Concerning Vessels which Have or Have Not Undergone Disinfection in View of

Destroying Rats and Mice on Board"), which was discussed by the Istanbul Quarantine Board over a period in 1901 and 1902. This originally
demanded ships arriving in Istanbul or other Ottoman ports to be treated
according to the sanitary conditions of the port of departure, regardless of
the health of the passengers and crew, and the existence or not of dead or
infected rats on board. However, in the course of its discussion, the Board
ended up adopting an even more draconian measure, which made both the
status of the port of origin and the sanitary status of the vessel irrelevant,
necessitating constant deratization.

The Ottoman circular stated that the following drastic and costly measures had to be applied in the quarantine stations and lazarettos of Ottoman ports in the Black Sea, the Aegean, across the Mediterranean, as well
as in Basra and the island of Kamaran to all vessels arriving from infected
ports. Merchandise, provisions, and any object on board was to be removed
from holds, underdocks, and the steward's room, so that baits consisting
in cheese or other foods that attract rodents may be placed therein. The
purpose of this was to attract rats from their nests and hiding places to the
open. Then furnaces would be placed at a rate of six per hold/underdeck
with two being additionally placed in the steward's room. Therein "sulphur
in sticks" would be burnt "in the proportion of 30 grams of sulphur to 60
grams charcoal for each cubic meter of space, or 30 kilos (66 lbs.) of sulphur to 60 kilos (132 lbs.) of charcoal for 1,000 cubic meters (35,317 cubic
feet) of space."[60] All spaces where this operation would take place would be
hermetically sealed and remain fumigated for at least ten hours. Alterna-
tively Pictolin could be used at a proportion of 1 kilo per 100m^3 applied for
three hours in a similarly sealed space. After reopening the holds, all rats
found suffocated therein would be thrown into the ship's boiler furnace or
otherwise burnt with the help of petroleum. The fumigated compartments
would then be washed with water, and its walls scrubbed with raw carbolic
acid, soap, and soda at 15 kilos each for 150 liters of water. As for other parts
of the ship (toilets, drawing rooms, cabins, etc.) and any object contained
therein, these would need to be disinfected by means of a 4 percent corrosive sublimate solution spray. Finally the water held in the ship's canteen
would need to be disinfected with lime.[61]

Expressing the dismay of the British Board of Trade about the Ottoman
circular, the UK Chamber of Shipping and the Liverpool Steamship Owners
Association, the British delegate to Istanbul's Quarantine Council, Frank

Clemow, proposed as an alternative the principles contained in the 1901 British memorandum on "Ship-Borne Rats and Plague."[62] This stated that only when plague is present on a ship or where, upon suspicion of plague, rats are examined and found positive, should quarantine and disinfection measures be imposed: "For a boat arriving from an infected port no measure whatsoever is obligatory: and only in the cases where we have encountered an exceptional mortality amongst rats in that ship is the destruction of rats, according to this Regulation, desirable."[63] As Clemow was quick to stress before members of the Quarantine Council, objections to the Ottoman approach were also voiced by the German Health Council, as well as by foreign shipping companies in Istanbul.

In an effort to reach some sort of compromise, Clemow crafted a list of internationally binding regulations on maritime sanitation, which, while assuming a draconian tone, in fact largely restated British positions on the subject. However, this effort to mitigate the stringent regulations imposed by the Ottomans backfired as it was usurped by the Ottoman authorities: "Where my proposal tended to grant facilities to navigation," Clemow wrote on October 12, 1902, "they have been rejected; where they tended to add fresh restrictive measures (as, for example, in regard to ships on board of which an exceptional rat-mortality has been observed) they have been accepted."[64] Clemow appeared desperate, for in his mind his proposal had aimed to "maintain the principles established by the various International Sanitary Conventions" so that, "while admitting that at the date of the last conference (1897) the part played by rats in the distribution of plague was not fully recognized and had therefore not been dealt with in the conference, I nevertheless maintained that the principles accepted by the Powers in dealing with the spread of infection applied equally well to rats as the medium of infection as to human beings, merchandise or ships."[65]

Indeed Clamow's proposed regulations encouraged the most draconian among the Istanbul Quarantine Council's members, with Greece's delegate, Dr. Balilis, rising to be the champion of even stricter measures in the course of the Council's sessions in September 1902, where Clemow's objections to the new regulations were discussed. Bringing up the case of *Polis Mytilini*, Balilis suggested that the clean versus infected port distinction was moot, and that the only way to prevent plague from spreading was to destroy all rats on all boats every forty days.[66] This followed Balilis's hitherto-stated thesis that plague was a disease residing in the soil, wherefrom rats and fleas got infected: In a pamphlet on the subject of quarantine authored the previous year, he had confidently expressed the opinion that it was from the infected soil lying in the crevices of boats that rats acquired plague, leading to an epizootic process that, by rendering plague more and more virulent through inter-rat transmission, finally produced the disease among humans.[67] Here the old hypothesis of plague's attenuation in the soil and its recrudescence through rats (see chapter 3) was rehearsed once again, but with severe practical consequences.

Balilis's lengthy exposition on sanitary measures against plague to the Istanbul Quarantine Council hailed the way in which the rat and flea connection led to the "unveiling of the mysterious agent which operates stealthily in the soil" by tracing the "hitherto invisible thread" of plague propagation.[68] By thus connecting soil, rat, and flea in a tripartite plague-spreading chain of transmission, "all that was until recently mysterious and obscure becomes knowledge and light."[69] In short, Balilis maintained that the prime locus of infection was the soil, with rats functioning as in-betweens the soil of different regions or countries, leading to the infection of "indigenous rodents" and eventually of humans, who may also contract plague directly from the soil.[70] Merchandise or luggage could also be infected likewise by the attenuated soil-borne saprophytic bacillus. As this, and not human contagion, was seen as the means of spreading the disease, quarantines needed to be reclassified as merely secondary means of prevention (eight days for the surveillance of humans who may be incubating the disease was deemed adequate) giving priority to deratization as the principle means of protection from the disease.

Boats, in Balilis's opinion, thus functioned not only as ideal transporters but also as vital reservoirs of plague. For if plague in its supposed soil-borne, saprophytic state was believed to wither and die when exposed to light or fresh air, in the holds of the boats both these conditions were absent, providing an excellent environment for the preservation of the disease in an attenuated state. If humans had nothing to fear from this attenuated bacillus directly, the recrudescence believed to be brought about by the rat, and rat epizootics in particular, was thought to lead to cases of the disease even on boats not deriving from an infected port for a long time, or even ones who had never been in touch with an infected port. "It is therefore the ship, this mobile soil, which is the thread that connects contaminated countries to unharmed countries, which must occupy our full attention."[71]

This theory effectively rendered the status of origin ports irrelevant, making the ship itself the main object of epidemiological interest and concern. This was an opinion that Clemow (who attached Balilis's pamphlet in his report to London) considered "illogical," "dangerous," and "subversive of all hitherto accepted principles and measures."[72] However, to Clemow's great annoyance, his European codelegates appeared largely uninterested in his proposals, and the Greek motion carried the day. With strong support from France's delegate it was adopted as an amendment (Article 3) of Circular No. 180 by the Quarantine Council, hence instituting the most draconian system of deratization on the globe. Clemow lamented:

> This article forbids clean ships from clean ports from coming to the quays, if they have touched a contaminated port within four months and have no certificate of rat-destruction less than forty days old. I took the example of a ship which leaves a contaminated port, say on June 1st: she has her rats destroyed before leaving and obtains a certificate to that effect and out of date. She goes to a number of clean ports during three and a half months, and is every where [sic] received without any restrictions. Bur she arrives in a Turkish port, say on September 15th, without undergoing the enormous expense and loss of time of a fresh rat-destruction, she is forbidden to come up to the quays.[73]

In light of this development in the doctrine of "perfect disinfection" (in Balilis's terms), fumigation apparatuses became crucial not only in practical but also in political terms.[74] No longer simply technologies that guaranteed the feasibility of continuous deratization, they now formed part of a broader interimperial struggle for ascertaining the correct and most "economic" approach for disentangling maritime trade from plague, and for securing a minimization, if not abolition, of cargo-related quarantine and the economic losses it entailed.

5 Stabilizing Fumigation

In the course of 1902 and the Ottoman deratization crisis, British health and port authorities followed closely the French experience with the Clayton process. Triggered by Frank Clemow's hesitation over the method, on January 14 the *British Medical Journal* received a letter by Clayton himself addressing the issue of the damaging of goods by his method. Whereas, he argued, this was true empirically speaking, it was only the result of the reaction of moisture with the chemicals produced by his machine; keeping merchandise protected from atmospheric moisture would guarantee no damage done to them by the Clayton gas.[1] At the same time, Clayton argued, his gas had a singular effect on rats: "Rats and all vermin seek to escape from the pungent gas as soon as they smell it, and its effects are slow enough to allow them to leave their hiding places before they succumb. This is not the case with carbonic gas of course, and serious trouble has frequently been caused by rats dying from its effects behind ceiling and panels, with quite offensive consequences."[2]

Still the efficacy of the Clayton machine continued to raise a range of doubts. A most telling conflict of opinions implicated Patrick Manson, who declared the Clayton to be "most efficient means of clearing ships of rats and vermin."[3] This endorsement elicited a vitriolic response by the chief medical adviser of the Local Government Board, W. S. Power, who bluntly declared: "I would that Dr Manson's knowledge of administration—of sanitary measures in their application in practice—were parallel with his enthusiasm in parasitology. The Board would be thus saved some unnecessary correspondence with the Colonial and Foreign Offices."[4] In Power's opinion, what others saw as minor or temporary misgivings of Claytonization (for example, "in one instance, there was slight bleaching of a dry sample of coloured silk") in fact necessitated new experiments on the process to be conducted in the United Kingdom.[5]

As a result, by invitation of the Clayton Company (June 1903), experiments were conducted at the Royal Albert Docks under direction of the well-known bacteriologist and cofounder of the School of Hygiene and Tropical Medicine, W. J. Simpson, in the presence of Prof. Hewlett and Dr. Woolston of Kings Collect and Sir William Hooper of the India Medical Council. The experiment, using plague and cholera cultures, was performed on the steamer *Manora* on empty holds. Reporting on the experiment, the assistant secretary of the Marine Department, Peter Sampson, related that within fifteen minutes two rats, which had been placed in cages within the hold, were retrieved "quite dead," adding that "their eyes were of a peculiar purple colour and I have some reason for believing that one of the effects of the gas is to act upon the eyes of living creatures."[6] Besides these experimental animals, around 220 rats living in the hold were also found dead. Rather than being conclusive, this apparent success did not suffice to convince British authorities of the appropriateness of the Clayton process, as the experiments had been conducted on empty holds and not with in situ cargo. Yet such hesitation seemed well past its hour in light of renewed trouble from Istanbul.

Just a few days after Simpson's experiments, notice came from the Ottoman Empire of Dr. Duca Pasha's proposal that vessels containing dead rats and deriving from plague-infected ports should be repulsed to the nearest lazaretto, where the rats would undergo bacteriological examination.[7] The Istanbul Quarantine Council was given but a week to consider the proposal, a diktat that infuriated the British due to its significant economic consequences: "It is an innovation which will bear very hardly on shipping and the sole justification is the present fashionable theory of the dangers of infection by rats, a theory which so far as I know reposes on conjecture merely and may tomorrow be discarded in favour of some new idea."[8]

Even if not everyone agreed with the author of the British response, E. C. Belch, that the rat's relation to plague was simply a "fashionable idea," relentless Ottoman insistence on maximum deratization urged for renewed efforts of experimentation with the Clayton apparatus, this time under the supervision of Franck Clemow, the British delegate to the Istanbul Quarantine Council.[9] The Clayton experiment, which took place in Liverpool on

July 30, 1903, was performed on the four-hold steamship Westmoreland of 1,384 tons gross register. Such was the haste that it was performed with only twenty-four-hours notice.[10] Only the two front holds were fumigated, one with a capacity of 14,000 cubic feet and the other with 22,000. Both were emptied of cargo before fumigation, with no rats having been recently observed onboard, possibly due to the vessel being used for the transportation of cement. The experiment aimed at ascertaining the germicidal power of the Clayton on cultures of anthrax, cholera, glanders, and enteric fever, (but, possibly on account of the hasty process, no plague), and its effect on "delicate food-products," including ship's biscuits, tea, coffee, sugar, apples, grapes, one banana, carrots, and skinned potatoes. Contained in test tubes, all bacterial cultures were wrapped and packaged in different ways and exposed in different parts of the two holds.[11]

The Clayton was then used so that the gas was pumped into the hold for four hours and fifty minutes, the gas being delivered at half-hourly intervals by the Company's engineer at a concentration ranging between 10 and 15 percent. After the engine was stopped, the hatches were kept shut for another three hours and then left opened for seven and a half hours when air was pumped into the holds for two hours. The food articles were then removed and exposed to the air in Clemow's private hotel room. While sugar, coffee, tea, and biscuits remained unaltered, bananas were "reduced to a semi-liquid, gelatinous pulp, which flowed through rents in the skin."[12] Grapes turned from black to green with an acidic flavor. As for apples, potatoes, and carrots, they all appeared to have softened as a result of the operation. At the same time, the bacterial samples were taken to Thompson-Yates laboratories, where no growth was observed in any of them, in contrast to the control tubes, where "there was a luxurious growth."[13]

Of particular interest was a peculiar aspect of the experiment, which involved rats that had been placed in the holds after being killed with typhoid bacilli. For this was the only case where the bacilli were not destroyed by the Clayton gas: "Cultures were made from the heart-blood of one of the rats, and a mixed growth of the typhoid (or enteric) bacillus and of the bacillus coli communis was obtained."[14] As a result, Clemow concluded that the Clayton gas had not penetrated the tissues of the cadavers. And still he remarked that this had failed to answer the main question of the experiment:

Firstly, I had wished to employ rats inoculated with plague, but as it was thought that there might be some risk in exposing plague-infected rats, typhoid-infected rats were employed instead. Secondly, I had wished to use living rats, in order to determine whether the gases, after killing such rats, would also destroy the infective material in their blood and tissues. But both rats died before the time of the experiment. It is quite possible that, had they been placed alive in the hold, and inhaled a quantity of the sulphurous gases before death, a sufficient amount of the gases might have been absorbed in their blood to destroy the infection there.[15]

Although Clemow did demonstrate that the Clayton gas had considerable penetrating power, the value of his findings was immediately challenged by his superiors: "the experiments here recounted do not help us much, as they were conducted on an empty vessel and not one with cargo 'in situ.' Until this is done our knowledge remains 'in status quo.'"[16]

The inconclusiveness of the experiments, in the eyes of British authorities at least, meant that any official endorsement or legislation on the preferred or sanctioned means of fumigation was stalled. What followed instead was a series of international conferences where the question of fumigation and deratization assumed central stage, setting the pace for the cross-European stabilization of these processes with regards to maritime trade. What was at stake in these meetings were the rules applying to quarantine and fumigation that port authorities of the participating states would have to adopt against the spread of infectious diseases while keeping the minimization of detention time as a generally accepted economic objective.

The Brussels Medical Congress

The first of these meetings was the Brussels Medical Congress, held in September 1903. There, drawing on his experience with plague in Arabia, Dr. Frédéric Borel, later known for his study of cholera and plague in relation to the Hajj, stressed the importance of focusing on the problem of the rat and maritime trade.[17] In order for plague prevention to be effective, he argued, a new international agreement on the classification of boats needed to be reached, in his opinion, in the following distinct categories:

1) A boat which, even if it derives from an infected port, has mortality neither amongst humans nor amongst rats. For this boat, immediate freedom must be accorded; 2) A boat that in course of its route has human cases without rat

mortality. We isolate the ill; as for the other passengers or the boat itself, we let them go free; 3) Finally, the boat which has mortality amongst rats, with or without human cases. This is the only one which is dangerous.[18]

To face the global peril posed by plague, Borel argued, it was necessary for port authorities to approach the destruction of rats on board of ships not as an exceptional measure but as what is "normal and regular. We need to destroy rats on board with the same regularity that we repaint a boat."[19] In other words, Borel propagated "a constant surveillance, a surveillance in all instances" of boats, which had "become veritable floating cities, entire cities, and cosmopolitan cities."[20]

On his part, the leading Pasteurian, Albert Calmette, also stressed the need to recognise the recently established importance of rats in the maritime propagation of plague, even when "there are no human cases onboard or the boat did not derive from a contaminated port." The goal that port authorities and the public health establishment should both and jointly aim for, Calmette argued, was for boat- and merchandise-related quarantine to be limited to "the time strictly necessary for the *destruction of rats and of insects* and in *complete disinfection of all the parts of the boat and the cargo.*"[21] To achieve this, what was required was the employment of scientific and officially sanctioned methods. Only thus, Calmette argued, could the aim of the progressive reduction of detention time and the eventual abolition of quarantine be ultimately reached.

As part of the discussions on the modification of prophylactic measures by port authorities against plague, and quarantine rules in particular, Calmette noted the limited efficacy of carbonic acid and carbon monoxide as fumigating substances that were both dangerous and did not kill the key objective of fumigation: rat fleas. By contrast, he stressed, an apparatus existed that combined sulphuric acid and SO_2, giving excellent results. This was no other than the Clayton machine. Stating that fumigation led by this apparatus was superior to the generally inefficient practice of quarantine, Calmette's optimism was countered by a more cautious approach by Germany's delegate, Nocht, who presented three objections to the use of the Clayton machine. First, in order to kill all rats on board of a ship it would be necessary for the relevant port authority to introduce sulphuric gas at a concentration of 10 percent, which Nocht doubted the Clayton could achieve. This was based on his own experiment of placing one hundred rats in cages in a boat's hold and surrounding them with different commodities—an experiment that did not result in the death of all the rats. Nocht's second objection resolved that sulphuric disinfection only had superficial effects and did not destroy all plague bacilli, such as those found in the excrements and corpses of rats. Finally, Nocht repeated the oft-voiced opinion that sulphuric gas had a deleterious effect on commodities, which, though not manifesting any exterior alteration, presented considerable changes when examined chemically, with tea, tobacco, and wool in particular accumulating great quantities of the gas.[22]

A staunch supporter of carbon-based solutions, Nocht also attacked Calmette's proposal to prove the efficacy of the Clayton by conducting experiments where tubes with plague bacilli would be placed inside goods in the hold of a boat. This, Nocht claimed, was both dangerous and unrealistic, insofar as "it is impossible to put the experimental tubes inside the deep parts of the boat, where one can find the rats."[23] As an alternative, he proposed the use of carbon monoxide pumped through the apparatus he developed and operated himself in the port of Hamburg.[24] Responding to Nocht, the veteran of fumigation experiments in Tunis, Langlois, stressed the need to distinguish between ordinary sulphurization of boats and "Claytonization," where the combination of SO_2+SO_3 "easily surpass 10%"—the disinfection of 45 boats in Dunkirk in this manner thus far had shown, in his experience, a total destruction of rats without altering or damaging merchandise, including fruit, tea, flour, and tobacco.[25]

Such deliberations, as they unfolded in the course of the Brussels Congress, were largely seen and acted out as preparatory discussions in anticipation of the International Sanitary Conference in Paris, which was to take place in December of the same year. However, the French, Pasteurian approach seemed to pre-empt this development insofar as only a few weeks after the conclusion of the Brussels proceedings, the President of France proceeded to issue a decree titled *The Destruction of Rats on Board of Ships Derived from Countries Contaminated with Plague, Before Unloading* (September 21, 1903). The decree stipulated that the destruction of rats from aforementioned countries was obligatory and that this should be affected by means of processes and apparatuses whose efficacy had been approved by the consultative committee of public hygiene of the Republic, and that it was immediately applicable in such harbors where these processes and apparatuses were in place.[26]

British Experiments

Faced with the embroiled debate in Brussels, the French presidential decree, and the impeding sanitary conference in Paris, British port authorities quickly moved to perform a range of new, more systematic experiments with the Clayton process. J. S. Haldane arrived in the port of Dunkirk on October 6, 1903, to witness the disinfection of *S.S. Bavaria*, a 3,006-ton cargo boat, by means of the Clayton machine. Haldane had inhabited a peculiar position in the history of hygiene, and his research and scientific practice were driven by a strong vision of social change.[27] In studies on the quality of air in densely populated parts of Dundee, in Scotland, Haldane had an interest in the toxicity of various gases, as he investigated their capacities to impair human environments. Furthermore, industry-based research in mine safety, the ventilation of tunnels, and poisoning by light-ing gas had, as Steve Sturdy has shown, "established him firmly as the lead-ing scientific expert on protective measures against noxious atmospheres in buildings and workplaces."[28]

The *S.S. Bavaria* had recently arrived in London from Calcutta, carrying cargo for the British capital, Glasgow, and Dunkirk. Having first made a stop in London, upon reaching Dunkirk its main bulk consisted of bales of jute, poppy seed, linseed, and pulse grain destined for the French harbor. The boat arrived the following day and without delay, within two hours of arrival, a Clayton apparatus was brought aside the vessel and disinfection commenced in the presence of J. B. Evans of the Clayton Company. First the machine was used to fumigate the boat's two storerooms, a process lasting thirty minutes. Haldane reported that "the whole space appeared to be pretty filled with gas. Ten rats and thousands of dead cockroaches were lying about the floors, &c., which appeared to have been very success-ful."[29] Following this, different compartments of the boat were treated in sequence. As these bore ventilators, "the gas was introduced by putting the hose down a ventilator, the return pipe being usually put through a partly-opened hatch, and the other ventilators closed with canvas."[30] An hour or so was spent fumigating each of the compartments until "the gas began to issue in a concentrated state from the hatch."[31] Without treating the engine room or stokeholds, the last compartment to be ventilated was the saloon. As by then dusk had fallen, electric lamps were turned on, and Haldane reported being able to "observe through the windows the behaviour of the gas in the saloon": "It gradually filled the room from below upwards, the upper surface of gas having a fairly well defined surface. Above this layer flies still alive could be observed until the upper level of gas reached them, when they were killed."[32]

Fostered by this eerie image, Haldane's interest was particularly held by the way gas functioned in relation to the cargo itself. Noting that the temperature of the bags of seed contained in the holds was about 100°F, Haldane observed that "the air would naturally circulate down the sides and upwards through the cargo; and the introduction of a heavy gas such as sulphurous acid, would increase this circulation."[33] However, as the cargo was tightly packed, the little time allowed for fumigation prohibited a thorough penetration of the cargo by the gas. This was a problem that chimed with other limitations that Haldane had observed in the process. Descending to the two holds of the boat, after their ventilators had been opened for several hours, he conducted a thorough examination that left much to be desired as regards the efficacy of the Clayton process, at least as operated in that instance. For upon reaching the holds, Haldane was met not with piles of rat corpses, as it had been hoped, but instead by the chirping of crickets, which had apparently survived fumigation, "doubtless because the treatment had been too hurried."[34] Worse even, live rats and mice were said to be discovered and killed post-fumigation by the boat's crew.

On the other hand, Haldane was satisfied to confirm the results of Drs. Duriau and David as regards the harmlessness of the process on the cargo. The French doctors were eager to show Haldane exposed and control samples, which "entirely bore out their published statements."[35] Only the brass fittings in the saloon appeared a bit tarnished; still this was but a superficial effect, easily removed. And yet, Haldane observed, the process left an "extremely unpleasant" odor on beddings and woollen materials:

> I slept in one of the staterooms after the disinfection, and although the bedding has been aired on deck, and all windows and doors left open for several hours, the atmosphere was most unpleasant. Two ladies who slept on board suffered considerably from catarrh of the air passages, caused by sulphurous acid given off from the bedding.[36]

In spite of constant airing, Haldane complained that his clothes continued to smell "very distinctly" even three weeks following fumigation.[37] This made him suspect that had fumigation lasted longer, for SO_2 to sufficiently penetrate the cargo, the seeds contained therein would also have acquired the unpleasant odor, rendering the unloading of the cargo a particularly unpleasant job. Haldane thus concluded that, while effective in empty holds, "there is no satisfactory evidence as yet that it is practicable, without considerable delay and inconvenience, to make it equally effective in holds which are full."[38]

Prompted by these results and needing to know more about how the Clayton operates and the degree of its efficacy or harm, the Medical Office of the Local Government Board instructed Dr. John Wade to conduct experiments with the machine so as to ascertain the "percentage of sulphur dioxide required to kill rats and insects in a reasonable time," "its action on pathogenic bacteria, especially those of plague," "the manner in which sulphur dioxide and 'Clayton gas' penetrate cargo," and, finally, "the damage, if any, likely to be done to the cargo and appointments of a ship during the process."[39] In order to conduct experiments with the Clayton machine, Wade hired an experimental plant on a wharf at Blackpool, England, from the Clayton Company. The plant, originally designed for demonstrations of the fire-extinguishing capacities of the machine was converted accordingly.

Wade gave an evocative, and indeed rather impressionistic, image of the Clayton in operation:

> When the generator was working freely the surplus gas streamed from under the caves as a white fog, which, on calm days, fell rapidly to the ground, and then spread slowly like mist. On windy days it was impossible to work except on the windward side of the shed, the gusts of gas ejected from the building rendering the air in the immediate neighbourhood quite unbearable.[40]

In a manner befitting long-standing traditions of medical heroics, Wide decided to determine the maximum percentage of endurable SO_2 by experimenting on himself: "After the shed had been fully charged with gas, the doors were opened, and as soon as the interior could be seen, I endeavoured to enter and take a sample of the air."[41] Wade had to wait until SO_2 concentration had fallen to 0.2 percent to enter the shed, and even then he soon had to exit the premises as "the gas was practically irrespirable": "[A] little later, when the sulphur dioxide had fallen to 0.1 per cent., the smarting of the eyes, although still painful, was no longer so acute as to interfere with vision, and the air could be breathed with some difficulty. It gave rise to marked bronchitis and nasal catarrh the next day, accompanied by severe headache."[42] Wade reasoned that while people like the generator's mechanic developed tolerance to the gas, its distinctive smell, allegedly "perceptible at 0.01 per cent," as well as the visibility of sulphur trioxide safeguarded the boat crew from the approaching gas.

In order to ascertain the effect of the Clayton on bacteria, Wade conducted experiments using strips of sterilized cigarette and filter paper and pellets of sterilized cotton-wool soaked in cultures of cholera, typhoid, plague, anthrax, and shiga. Exposing such cultures to different percentages of SO_2 and for differing time periods, Wade took care so that they were wrapped, thus simulating the practical conditions in the hold of a ship. In this way, his trial replicated Calmette's original germicidal experiments on the *René*. Each experimental "envelope" contained a full bacterial set, and was further "wrapped up in several thicknesses of cotton wadding, and then made into a small parcel with wrapping-paper. Finally, each parcel was wrapped in several thicknesses of blanket, making a bundle about nine inches long as a little less in diameter."[43] Indeed, Wade went so far as to place two of the wrapped packages in a locked steel trunk filled up with clothes or blankets. The key question of these experiments was to test the Clayton Company's claim that the gas generated by Claytonization "is more efficient as bactericide than pure sulphur dioxide diluted with air."[44] Wade concluded that his experiments demonstrated that, whereas in the case of plague such difference was negligible, "a slight but distinct superiority in the toxic power of the Clayton gas as regards typhoid germs" was evident; this he attributed tacitly to the dampness of the material exposed to the gas.[45]

Experiments on rats performed by Dr. M. S. Pembrey, a physiology lecturer at Guy's Hospital, London, had already shown that the maximum percentage of SO_2 tolerated by rats without incurring death was around 0.2 percent, with young rats being more susceptible to the gas. Wade concluded that "the uniform diffusion of 0.5 cent. of sulphur dioxide through the hold will kill the rats within 2 hours" and that, as a result, the Clayton Company's recommendations for six to seven hours of exposure with 0.10–0.12 percent SO_2 was sufficient, with the provision, however, "*that the gas penetrates into every part of the cargo.*"[46]

These results led him to conclude that while effective on an empty hold, the Clayton process faced distinct limitations when faced with holds bearing densely packed cargo. This, he reasoned, was due to the propensity of the gas to be absorbed by material, especially wool, tea, coffee powder, cocoa, and flour. In the case of wool, which he considered a prime locus of SO_2 absorption, Wade claimed that "a bale of compressed wool would absorb at least 10 times its volume of the undiluted gas," which it would then give up upon exposure to air.[47] Given this extensive absorption, the penetration of the gas into the totality of the loaded hold required what Wade thought was a very long time. To ascertain the precise span of the latter, further experiments were conducted with bales of wool provided by the Clayton Company, which also gave technical support in constructing an experimental apparatus through which gas could be inserted directly into the bales. After several experiments with wool and jute bales, Wade concluded that, no matter how tight the packaging, SO_2 was indeed able to penetrate this material with such rapidity and to such extent that there is no "possibility of distributing throughout a [jute or wool] loaded hold a percentage of sulphur dioxide to kill every rat," which could easily find refuge from the gas between bales and sacks.[48] On the other hand, no such limitation was shown to apply to cargoes of cotton, which bore none of the absorbent qualities of the other material. In conclusion, Wade claimed that when it came to empty holds, all rats and insects could be destroyed by means of a fumigation with 0.5 percent of SO_2 over a period of less than two hours—a result, however, untenable in loaded cargo conditions due to the high and rapid absorbability of the gas. Among different bacteria, plague was deemed optimally destroyed at a concentration of 2 percent over a six-hour exposure, while typhoid, being more resistant, required twenty-four hours under the same concentration. Confirming Robert Koch's older experiments, anthrax, on the other hand, was deemed to be immune to SO_2.

As regards the question of damage caused by Claytonization to cargo, Wade noted that the dampness observed post-treatment in textile materials and fabrics was radically reduced if these were suspended vertically, or covered even by a single layer of thin tissue paper. Similar protection was afforded in this manner against the damaging effects of SO_2. In any case, even when goods were permanently damaged, no residual odor was observed, contradicting Haldane's report, which Wade attributed to the probable presence of small quantities of selenium in the gas. As regards foodstuffs, Wade confirmed that tobacco, fresh fruit, and vegetables, as well as powders (but not sugar or salt) were among those critically impaired. Of particular interest to him was flour, which not only seemed to be losing its odor and taste as a result of fumigation, but also evinced reduced ability to "raise," thus making it useless for breadmaking.

Faced with these results, in their joint follow-up report, Haldane and Wade sought to compare the Clayton process with other processes of rat destruction in ships. First, they admitted that the Clayton method "possesses very distinct advantages": markedly, the fact that Claytonization worked by means of the law of gravity, with the gas finding its way to the bottom of the holds due to being heavier than air. At the same time as obviating any risk of fire or explosion, and having a very low risk of asphyxiation, the use of SO_2 was deemed superior to carbon monoxide, both in terms of its visibility to the naked eye and in its ability to kill insects. Still, if the Clayton apparatus was deemed to provide an efficient method of disinfection, this was conditioned upon the penetration rate in practice. Hence, in light of Wade's findings, the report expressed its reservation as regards the actual, practical efficacy of the machine as applied to loaded ships. The methods compared to the Clayton were the following:

a) Burning sulphur on shipboard, which could only be used on empty holds and was deemed to be effective against plague in particular, but a tedious task involving a high risk of fire.

b) Liquid sulphurous acid, employable in both empty and loaded holds, which was deemed somewhat less damaging than Claytonization yet not obviating the latter's absorption/penetration problems.

c) Carbon monoxide, which, having already received the attention of the Medical Officer of the Local Government Board and doctors like Nocht, was being currently used for rat destruction on loaded vessels in Hamburg—a cheap and easy-to-use substance, CO's advantage was that it was not absorbed by cargo. On the other hand, the substance was dangerous to humans due to being odorless and, under certain conditions, explosive, while having no disinfectant or insecticidal action. In Hamburg, an apparatus was used so as to obviate such problems by transforming CO to CO_2.

d) Carbonic acid, which was safer but required large quantities to kill rats while not killing insects.

It would be thus fair to say that, in Wade and Haldane's view, circumstantial factors benefitted different fumigation methods:

If all vessels from plague-infected ports, whether known to be infected or not, were to be treated, the question of expense of materials and of the damage producible by the sulphur process would be a very serious one, and probably the carbonic oxide process would be preferable on the while in cases of known infection of vessels, the treatment with carbonic oxide would doubtless need to be followed by a separate disinfection process. The question of a certain amount of damage by the sulphur process on vessels actually infected with plague, of which there are likely to be very few, is of no great importance, particularly as disinfection will be needed in any case. In view of all the circumstances we think, therefore, that a sulphur dioxide process will be most generally useful.[49]

If, as a result, Wade and Haldane advocated the use of the Clayton as particularly useful in large English ports, doubt still lingered regarding the efficacy of the method on loaded cargo, and more experience was deemed necessary so as to ascertain the optimal duration of the treatment. In case this proved to be too long, then an option would be to add carbon monoxide (aka carbonic oxide) to the gas in the holds. In that case, it was suggested "The diminished proportion of oxygen in the interior of the cargo treated by the Clayton process would greatly increase the poisonous action of the carbonic oxide on the rats, which could thus be destroyed with certainty, while at any rate a certain proportion of the cargo would also be disinfected."[50]

Held only weeks after the conclusion of the British experiments with the Clayton, the sanitary conference in Paris promised to be the decisive moment for an international decision on the optimal fumigation method in the quest of a quarantine-free future for maritime trade. As had often been the case, however, rather than reaching a consensus, the international meeting only came to confirm scientific discord.

The Paris International Sanitary Conference (1903)

Coming together for the first time since the 1897 International Sanitary Conference in Venice, where international maritime quarantine rules had been discussed and defined, twenty-five nations gathered in Paris in December 1903. This was the first legislating international sanitary meeting to tackle the case of the rat's implication in the spread of plague. Indeed, the person internationally recognized as the canonical summarizer of the 1897 sanitary conference, Adrien Proust, stressed that while the purpose of the 1903 conference was to codify existing regulations, as passed in 1897, the discovery of the role of the rat and its flea in the propagation of plague had led to the need to discuss the effective measures of vector destruction, or even germicide, on board of ships.

The hope was that such measures might be able to rid vessels of diseases like plague, yellow fever, and cholera, thus bringing about an immense advantage for international sea trade. Yet, as expressed by the French delegation, this required a radical revision, in light of recent scientific developments, of the definition of "unaffected" boats (*navires indemnes*): "In Venice, in 1897, we only concerned ourselves with ships from the still contaminated or suspect regions of the Far East. The term unaffected [*indemne*] was applied to ships which, although originating from contaminated areas, had had no sick persons during the voyage."[51]

As of recently, however, the role of the rat on maritime plague had been connected with observations that plague could break out onboard a ship even up to sixty days after it had left a contaminated harbor, making the disease's incubation period in humans irrelevant. Indeed the establishment of the rat as a key player in the maritime propagation of plague exposed a paradox, aptly embodied in the key seafaring crossing of the Suez: "A vessel, conveying a patient isolated on board, and whose presence is almost no longer dangerous, will be quarantined. A vessel infected with plague-rats, which offers much more peril may freely pass the Canal."[52] As a result, and in order to bring back to the notion *indemne* its true meaning, Proust proposed the following redefinitions:

A. As *unaffected,* the ship from an uncontaminated district which has not stopped at contaminated ports, which has not had a confirmed or suspected case of cholera, yellow fever or plague, and on which no plague-infected rats have been found.

B. As *suspect,* the ship from a contaminated origin, or having stopped in contaminated ports, but which has neither confirmed nor suspected cases of cholera, yellow fever or plague, and on board of which the presence of plague-infected rats has not been observed.

C. As *infected,* the ship from a contaminated or uncontrolled district which has stopped at contaminated or uncontaminated ports and which has presented confirmed or suspected cases of cholera, yellow fever or plague or on which plague-rats have been found or observed from the time of departure or arrival.[53]

For a port to be considered as no longer contaminated, a bacteriological examination of rats living therein was deemed necessary; only when these proved to be free from plague could a hitherto infected port be declared unaffected. But what was the role of rat destruction in this scheme of things? Proust harbored visions of deratization apparatuses being installed in every port—a duty that, in the eyes of many of his colleagues, should burden ship companies and not the state and its port authorities. As the Belgian delegate's statement demonstrates, the vision entailed here was uniquely utopian, pointing not simply to the limitation but to the abolition of quarantine: "In order to obtain progressively the reduction of the duration and even, if possible, the total abolition of quarantines, encourage shipping companies and ship-owners to complete the destruction of rats and insects on board their ships after each complete unloading of the cargo holds under the control of sanitary administration."[54] Upon deliberation, the "technical Commission" of the conference thus agreed on a new classification:

> A ship which has plague on board or has had one or more plague cases for seven days is to be considered *infected*. A ship which there have been cases of plague at the time of departure or during the voyage, but no new cases have occurred during seven days, is to be considered *suspect*. A ship which has had neither dead nor cases of plague on board, both before departure and during the voyage or at the time of arrival, is to be considered as unaffected, even if it arrives from a contaminated port. Ships aboard which the presence of plague-infected rats is or has been in place, shall be subjected to special measures.[55]

Still what remained a contentious issue was whether the observation of a rat epizootic in a given harbor should be made notifiable in the same way as human cases were. The position was defended by Proust in opposition to a broad consensus that epizootics should be noted only in case of human outbreaks. The reason for this was that Proust's proposed regulation would be difficult to implement and would, as a result, favor countries that simply did not conduct tests on rats, thus potentially raising interstate suspicion that could be detrimental to trade.

At the same time, in light of the centrality of the rat, a heated debate unfolded regarding the optimal method of deratization. While the leading Pasteurian, Albert Calmette, argued that the conference did not need to decide over one process or another, and that each country should be free to decide for itself, he still insisted that from among the three proposed deratization methods (H_2CO_3, SO_2, CO_2), only one had disinfecting qualities: the Clayton. This direct endorsement was, however, immediately met by strong opposition led by the German delegate and associate of Nocht (better known as the discoverer of the causative agent of typhoid fever, *salmonella typhi*), Georg Theodor August Gaffky. Recalling Robert Koch's old anthrax experiments, Gaffky maintained that the aim of these processes was deratization and not disinfection, casting doubt on the germicidal properties of sulphuric acid. Moreover, pointing at Nocht's opinion that sulphuric acid damages the boat's metal plates, Gaffky stressed that the effects of these gases on goods had not been adequately established.

German opposition only came to underline a broader confusion, as Calmette was quick to point out: the mistaken equation of the Clayton gas with sulphuric acid, H_2SO_4, when in fact the former was sulphur dioxide, or SO_2. Upon Calmette's question on the subject, Nocht, who was also present at the meeting, specified that indeed his own experiments had been with sulphuric acid and not the Clayton gas, still maintaining, however, that the same deleterious effects would probably result from the latter. Calmette was quick to explain that while open-air production of sulphuric acid would lead to such damage, the production of the gas inside the Clayton apparatus left surfaces unharmed. He stressed that any confusion between "the gas produced by the Clayton apparatus and that results from the open air combustion of sulphuric acid" should be avoided, "the former having, because of the special conditions in which it is obtained, a much higher concentration to which it owes an incomparably superior disinfecting power. While its content reaches 9, 10, 12 and even 1 p. 100, that of the gas produced by combustion in the open air does not exceed 3 to 4 per cent."[56]

Nocht, however, insisted in his attack on the Clayton, this time from a financial point of view, arguing that Calmette had himself said that "flour impregnated with sulphurous acid became unfit for bread-making, but that it was sufficient to aerate it to restore its properties."[57] How, Nocht objected, would this be possible in the case of flour contained in barrels or bags, without "removing from the recipients the guarantees resulting from the trademark of those products"? Nocht stressed that, as a result, "in Germany the use of sulphurous acid would certainly give rise to claims which would result for the State by the payment of heavy indemnities."[58] The German opposition was to find an unexpected ally in the British delegate who, in a move that relied on a selective reading of the experiments conducted by Haldane and Wade over the previous months, stressed the inconclusive nature of experiments regarding the time required for the Clayton to bring about the complete deratization of ship holds.

To Calmette's support came Proust who claimed that, since its application in Dunkirk, forty-five boats had so far been thus fumigated with no recorded damage. Proust distributed the report he coauthored with Faivre to the delegates. He and other French delegates proposed, in light of the different available methods, that the conference should adopt the following formula: "In the present state of science, the means which have been recommended as effective are Clayton gas, carbon monoxide, and carbonic acid." However, Gaffky objected that this would appear like an endorsement of the first, when the deleterious effects of sulphuric acid were in fact in question. Governments, he suggested, should just be allowed to choose whatever method they wanted.[59] With delegates from different countries quickly taking sides along the Franco-German divide, the chairman of the Commission proposed the following compromise:

> In the present state of science, the means which have been recommended as the most efficacious for the destruction of rats on board ships are:
>
> 1. A mixture of sulphurous acid and sulphuric anhydride, propelled under pressure in the holds, and assuring the stirring of the air, together with a concentration of sulphurous acid equal to at least 8 per cent. 100 per cubic meter of air.
> 2. A mixture of carbon monoxide and carbonic acid.
> 3. Carbonic acid.
>
> It is advisable to organise in the ports a scientifically rigorous control, making it possible to ensure, for each disinfection operation, that the rats and the insects have been completely destroyed.[60]

But even this formula was rejected as unacceptable, as the British delegate claimed that it misinformed insofar as "it presupposes an acknowledgment of efficiency," while Gaffky, doubting the role of fleas in the spread of plague, suggested the removal of any reference to insects.[61] As a result, the closing paragraph of the statement was thus revised: "It is necessary to organise in the ports a check to ensure, for each disinfection operation, that the rats were completely destroyed. This control is not required when the operation is carried out by the health service," and was unanimously adopted.[62]

As a result of the acrimonious divide, delegates at the Paris Sanitary Conference proved unable to come to a meaningful consensus regarding a single optimal method of vessel fumigation. However, the conference may indeed be said to have achieved a stabilization of maritime fumigation. For this was the first time that this process, irrespective of the method advanced by different parties, was endorsed at the political and legislative level afforded by the conference. Rather than simply remaining an international aspiration, maritime fumigation had, by means of the 1903 International Sanitary Conference, become an internationally stabilized legislative and practical field. The impact of this would soon be felt across the globe.

Globalizing the Clayton

On May 4, 1906, following the advice of the Paris Sanitary Conference, a Presidential decree made the destruction of rats on all French ports obligatory before unloading of cargo for vessels arriving from or having touched at plague-contaminated ports, as well as for vessels having received via transfer from another ship merchandise exceeding fifty tons from a contaminated country.[63] Most importantly, according to the decree, the destruction of rats should be carried out "exclusively by an apparatus the efficiency of which has been recognized by the Superior Council of Hygiene."[64] Besides the Clayton, in the course of the following years these would include the Marot (from 1905), an apparatus using liquid SO_2 (see chapter 6), and the Gauthier-Deglos apparatus (from 1907), a machine employing sulphur and coal dust with the help of a ventilator that, by sucking out the air in the holds, forced the introduction of the gas into the vessel.

Besides vessels that had not touched upon plague-infected ports, also exempt from these measures were declared such vessels "that only land passengers in French ports without docking and which sojourn only several hours," and vessels that stop at French ports for less than twelve hours and discharge less than 500 tons of cargo (as long as rat-guarding measures are observed).[65] However, the decree also included in this exempted category more ambiguous cases, such as vessels that had travelled for sixty days since departing from a contaminated port without touching any other port in the meantime and where "nothing of a suspicious sanitary nature" had been observed onboard.[66] Similarly exempt were vessels that, in spite of having touched upon a contaminated port, could "prove that they neither

berthed alongside the quay or landing stages, nor embarked merchandise,"
as well as vessels of a similar origin whose captain could certify that it had
undergone deratization at departure following processes approved by the
French sanitary authorities.[67] According to the French decree, obligatory
deratization concerned holds, bunkers, and crew and emigrant quarters,
as well as third- and fourth-class compartments, but excluded first- and
second-class compartments and officers' cabins, as well as salons and din-
ing rooms. The cost of the operation was to be borne by the shipowners, as
already established by law since 1896.[68]

Outside France, the period following the 1903 International Sanitary
Conference witnessed the adoption of the Clayton in many ports, occasionally even in the service of deratization campaigns on land. Indeed
the decades following 1903 mark a peak in global preoccupation with rats
and deratization on both sea and land. What fueled this was a growing
international antagonism over hygienic modernity, with claims of a lack
thereof by opposing nations or empires functioning as an important tool
in maritime trade competition.[69] This is not to say that deratization efforts
were cynical. Far from it, all evidence points to the fact that it was a goal
seriously aspired to by public health as well as port authorities, reaching
such climaxes as the mass campaign launched in Denmark for the total
eradication of the species. There, calls for a legislation for the extermination
of the rat in the Danish Kingdom had been in place since 1898, but were
only enacted through the signing of a law by the Danish King in March
1907, following an aggressive campaign by the president of the Society for
the Destruction of Rats, Emil Zuschlag.[70] The national war on the rat led to
massive efforts of extermination; the grand total of rats caught and handed
in for a reward between the start of the campaign in July 1907 until January
1909 being no less than 1,557,656, not including rats destroyed directly by
the government.[71] It was an event that elicited global fascination and contributed considerably to the utopian fantasy of rat-free harbors, cities, and
even nations. If efforts to emulate the Danish example proved less than successful, they still marked a pattern emerging in Western Europe and North
America, while also affecting the colonies: rendering deratization a civic
duty.[72] Most importantly, when it comes to the global war against the rat in
terra firma, the deratization hype accompanied and fueled, both financially
and affectively, a surge for "building out the rat."[73] As a result of a widespread agreement that fumigation of buildings was not effective, due to the

manifest difficulty in sealing the latter effectively and to the danger posed to humans, while such practices were not altogether abandoned, they rapidly gave their place to rat-proofing as the catchword of urban hygienic reform and modernity across the globe.[74]

In the field of maritime sanitation, in the years following the 1903 International Sanitary Conference, the Clayton machine was widely adopted across the globe.[75] This included, for example, the Ottoman purchase of a Clayton machine for use in the key port of Jeddah in the spring of 1909, and the application of the Clayton in the port of Amoy the same year.[76] Similarly the Clayton was employed in the port of Trieste for vessels arriving from the Far East.[77] In Peru, the adoption of the Clayton machine was made compulsory by the country's government, so that after 1903 three sanitary stations became operative—in Ilo, Callao, and Paita respectively. Brazil in turn imported various machines that were brought into practice both in the port of Rio de Janeiro as well as on its streets.[78] All operated the Clayton machine on vessels derived from infected ports. But this was not enough. The Peruvian government also made it compulsory to install and operate Clayton machines on passenger steamers, with Pacific Steam Navigation Co., Cia Sud de Vapores and W. R. Grace & Co. complying with the decree. The American consul in Callao contrasted the impressive onboard results of this method to the disappointing failure of applying methods such as poisoning, trapping, and using pathogens, such as the Danyz virus, against rats in the harbor wharfs and buildings of the Peruvian port.[79] This global adoption of the Clayton was not, however, a frictionless process. Indeed, as we can see from a brief examination of the case of French Indochina, it involved pressing financial questions.

The application of the Clayton proved to be desirable across the French colonies, but in particular in Indochina, where plague had become a recurring problem since the first outbreak of the disease in the coastal town of Nha Trang in 1898.[80] In August 1908, following his meeting with the director of the Institut Pasteur, Émile Roux, the newly appointed Governor General of Indochina (GGI), Antony Wladislas Klobukowski, urged the superior council of hygiene of the colony to take every measure so as to adopt the most practical apparatuses for disinfection against plague. Responding to the GGI's question on whether Clayton-driven sulphurization of all cargo suspect of harboring rats and insects was in place, the council informed him that this was indeed already effective in Saigon, where it was applied when merchandise arrived from infected ports, especially Hong Kong during the summer months, when the Crown Colony suffered from its seasonal outbreaks of the disease.[81] However, the GGI and his agriculture director insisted on the need to sulphurize not only cargo ships but also junks (traditional Chinese sail boats), as these, in their opinion, posed an even greater threat than international steamships.[82] In practice, the key problem to this expansion of the use of the Clayton was financial. The cost of a type-B Clayton apparatus, suitable for such small vessels and deemed to be desirable for Nha Trang's port, was estimated to be at least 25,000 francs at the time— something that, the GGI admitted, Annam's budget could not afford. Could this then be considered an expense for the colony's general rather than regional or local budget? This faced the opposition of the Director General of Finances in the colony, who objected that such expenses did not fall under the umbrella of "general public health" and yet accepted for such purchase to be listed under the general budget expenses of local budgets, hence being indirectly funded by the central colonial budget.[83] Soon enough, at the session of the council on October 28, 1908, the general employment of the Clayton for the defense of the colony against plague was decided. The following were deemed necessary: two type-D Claytons (one on rails, the other riverine) for Hanoi, and two type-M Claytons for the border posts of Moncay and Ha-Giang, in Tonkin; one type-A Clayton for the maritime station of Tourane, and three type-D Claytons for Phantiet, Hue, and Bangoi, the last two on rails, in Annam; one type-D Clayton for Cholon and one type-M for Mytho in Conchinchine; one type-A Clayton for the maritime station of Pnom-Penh and one type-M for Battambang in Cambodia; one type-D apparatus for the riverine station of Khone, in Laos; and one type-M apparatus for the Fort-Bayard station in the French enclave of Guangzhouwan, in China.[84]

We can see here, in a nutshell, that the global expansion of the Clayton depended on a key design feature of the apparatus: its availability in different forms and sizes so as to fit different needs and aims. And yet, if during the years following the 1903 International Sanitary Conference, the Clayton had consolidated its global spread, by the turn of the first decade of the twentieth century it could no longer claim a monopoly of mechanized sulphur-based fumigation. A number of other apparatuses had started making their appearance in the global market, leading the way to a proliferation of different sulphur-based procedures and machines. Among them was a French invention, which promised an innovative electrification of SO_2 to increase its capacity for the purpose of disinfestation and deratization. Developed and tested in France, the Aparat Marot found its most expansive application in Argentina, where the *higienistas* aimed to transfer the hygienic model city of Buenos Aires into the twentieth century.

6 Fumigating the Nation—the Sulfurozador in Argentina

By 1906, Clayton's apparatus had been met with serious competition from a new French invention, developed to improve the security of French harbors around the globe. However, as this chapter will show, the new mobility and cost-effectiveness of the Aparato Marot catalyzed a sulphuric utopia of an urban space without pests. The Sulfurozador, as it would come to be known in Argentina, transferred decades of experimental maritime sanitation to urban space. We take the city of Buenos Aires here as an example to demonstrate how the disinfestation with sulphuric acid gas assumed a pivotal role in the complex fabric of social, moral, and racial politics of hygienic modernity at turn-of-the-century Latin America.

For Argentina's powerful medico-political elite—the "higienistas"—the turn of the twentieth century was a time of enthusiastic innovation, rapid progress, and growing hope to lay the foundation for a robust and healthy future for the young nation. Among the architects of this modernized utopian vision for a sanitary state was José Penna, a prominent physician who held the country's first academic chair of epidemiology in 1900. When he was tasked with controlling the capital's public health service in 1906, his most prestigious acquisition was a disinfection machine: the Aparato Marot. Invented in 1904 in France, six of these machines were imported and installed on horse carriages and automobiles to carry out comprehensive disinfection of homes, warehouses, streets, and the city's new sewage system. Eventually, as the rat became the health authority's key adversary, the machine's immense capacity enabled a comprehensive sanitation campaign, aimed ambitiously to disinfect and to disinfest the entirety of the terrain of Buenos Aires.

The apparatus yielded outstanding results. By injecting sulphur dioxide into enclosed buildings, both residential and commercial spaces could quickly be freed of possible sources of infection. The Marot came equipped with an additional innovative component. The device electrified the sulphur dioxide before propelling it out of the pipes, which created an ozone effect, resulting in a lighter gas. Thus, the gas could easily fill large rooms without extensive isolation, killing pathogens, insects, and rodents throughout the space. It was especially effective in annihilating rats often found climbing the beams of buildings to escape the lethal fumes. The machine, placed at the heart of Penna's ambitious scientific urban sanitation program, quickly became known to Argentineans as the Sulfurozador.

This local history of the Sulfurozador's relation to the twentieth-century hygienic vision of Buenos Aires illuminates and further exemplifies the larger transnational, perhaps global, push to establish rigorous and scientifically proven sanitation methods, outlined in the previous chapters.[1] Across Latin America, hygiene was seen as a driver for social change; public health programs had achieved a prominent status in late nineteenth-century state-building processes, combining urbanization, centralization, and the negotiation of individual rights within the public sphere. The national identities of Argentina, Peru, Brazil, and many other thriving nations was strongly influenced—as Marcos Cueto and Steven Palmer have recently discussed in detail—by the installation of institutions devoted to national health.[2] Routines, laws, and a morality of hygiene were adapted as "means of articulating political concerns in technical terms," both on national and (toward the end of the nineteenth century) on transnational levels.[3] Nation-building and the emergence of a field of state-governed medicine were, as José Amador has argued, deeply intertwined: "public health transformed more than ideas about disease," he writes. "[I]t entered public consciousness to shape attitudes about race, plantation life, urbanization, spaces of sociability, and personal responsibility."[4]

At the end of the nineteenth century, fumigation with sulphur dioxide had also become a central pillar of Latin American maritime sanitation. On the heels of a devastating cholera outbreak in 1886, an agreement was signed by Brazil, Uruguay, and Argentina in 1887, later joined by Paraguay, to establish sanitary conventions in all major South American ports.[5] A transnational scientific commission strongly recommended widespread fumigation of vessels with sulphur dioxide to reestablish trust in the trade connections across borders.[6] But it would take until the beginning of the twentieth century and the arrival of bubonic plague in Latin America for the new urgency for sanitary hygiene to be met with industrial and techno-logical innovation.[7] Across the continent—as well as in many other ports across the world—advanced fumigation machines such as the Clayton apparatus or the Aparato Marot were put in place to curb the transmission of yellow fever, typhoid fever, cholera, and bubonic plague on the path-ways of maritime commerce. But occasionally, as in Buenos Aires, but also in Rio de Janeiro, New Orleans, and San Francisco, these machines were brought into the streets to enable extensive disinfection campaigns with sulphur dioxide in the urban environment.

However, only in Buenos Aires did the introduction of a disinfection machine to the streets and homes of its residents lead to the resurgence of a utopian vision of total urban disinfection. We ask here, if and to what extent this vision was encapsulated in and catalyzed by the technological capaci-ties of the Sulfurozador. Many scholars have addressed Argentina's hygienic movement between 1880 and 1910 as a characteristic social and politi-cal foundation of public health in Latin America. But while the motives, traditions, and discourses of the higienistas have received much scholarly attention, their successful practices and the material changes their visions brought about, remain at times overshadowed.[8] Questions remain about how utopian visions of a hygienic future were implemented on Argenti-na's "epidemic streets" and how the expansive system of prevention poli-cies that would come to characterise twentieth-century Buenos Aires were materialized.[9] As this history is also marked by significant epistemological transformations, we ask here how the "intrusive interventions" of sanitary brigades were modernized and industrialized to transfer the hygienic utopia of the Argentinean capital into the twentieth century.[10]

In 1906, the Marot crowned a long series of innovative sanitary improve-ments of Buenos Aires, of which the most significant was the introduction of sanitary brigades. As Adriana Alvarez and Susana Belmartino have argued, systematic centralization and institutionalization in Argentina's health administration encouraged rapid introduction of modern methods.[11] After the national Department of Hygiene was founded in 1880, followed by the Asistencia Pública [public service] in 1883, the city saw the first disinfection brigades operating on the streets by 1888. As Alvarez emphasizes, the pur-pose of the brigades and their intrusive campaigns was not only to disinfect "pathogenic" houses, but also to demonstrate the impotence of traditional methods, which relied on old theories about miasma, still practiced mostly in private.[12] Moreover, historian Diego Armus has argued that the disinfection brigades effectively mobilized bacteriology to overtake miasmatic theories of infection, marking how this new science left the spaces of laboratories and hospitals to be used in a public and visible way.[13]

It was the arrival of bubonic plague in 1899/1900 that exposed the failure of existing sanitary strategies and concepts.[14] The outbreak led to a renewed push for improvement of the city's hygienic state and, as the rat moved to the center of sanitary campaigns, it encouraged a renewed sense of the city's environment as an infected terrain. In the aftermath of the plague, the Aparato Marot came to stand for two decisive transformations in the fight against infectious diseases in Buenos Aires from 1906 onward. First, its application in streets and houses symbolized a shift in the consideration of the origin of epidemics as the Sulfurozador should protect the urban environment against infectious diseases that were increasingly seen as endemic threats from within. Second, the Sulfurozador was key to the modernization of a long-standing innovative public health tradition, rooted in Guillermo Rawson's 1870s utopian vision of a sanitary state at the southern tip of Latin America.

The city's terrain, its streets, houses, sewage systems, and open land had once been the exclusive focus of old sanitarians' attention. With the Aparato Marot, we argue, this terrain became subjected to modernized, sophisticated, and experimentally proven disinfection. The machine mobilized maritime fumigation beyond the port to renew the visionary and popular poetics of a "Higienismo Argentino" during a period of growing positivism and rationalism.[15] Alejandro Kohl has described these utopian poetics to be drivers of the original hygienic program of the sanitarians Rawson and Wilde. They had merged since the 1870s the vision of an ideal society with campaigns to remove miasma, foul stench, and noxious vapors.[16] This legacy of "civic health" was translated now into a modern principle of "general prophylaxis." The integration of the new technology in 1906 symbolized the successful adaptation of traditional sanitary practices within the epistemological environment of modern bacteriology and laboratory science. We will here demonstrate how in technological descriptions of the machine, in the experimental evaluation of its capacities, and in the admiration for its all-encompassing disinfection qualities, a sense of the utopian visions of the old sanitarians was captured and renewed to envision a new sense of hygienic invulnerability.

We begin with a short overview of the hygienic movement of Argentina and the efforts undertaken to transform Buenos Aires into a model hygienic city since 1870. The second section illuminates the emergence and institutionalization of disinfection practices and focuses on the conceptual integration of bacteriological perspectives into Argentina's epidemiology. The third section shows how the unexpected outbreak of bubonic plague in Buenos Aires fueled the rapid development of novel strategies for sanitation, and foregrounds the rat as a new adversary for public health intervention. We then turn to the Aparato Marot, its descriptions, critical and experimental appraisal, and its elevation into a spearhead of the newly popular promise of "general prophylaxis." Throughout the chapter, we follow Penna, the key figure behind the introduction of the Marot, as he set up experiments to determine the capacity of these machines, the density of the chemical solutions required, and the manual routines needed to achieve complete eradications of germs, insects, and any potential animal vectors of disease, before we close with the celebration of the machine's triumph by his successors.[17]

Making Buenos Aires the Hygienic Model City

Since the 1870s, Buenos Aires experienced unexpected growth, which, according to Kohl, contributed to the city's newfound status as a laboratory for social and political visions. A new emerging "poetics of wellbeing" found immediate application in urban planning. The question was how to devise conditions of social life that would allow the population to conserve its health, keeping infectious diseases from entering the capital, while crafting a vision of the ideal state.[18] But the extensive programs, designed to secure the capital's hygienic status, also required rigorous sanitary policing.

Since the end of the Argentinean civil war in 1880, Argentina had seen the rise of a medico-political elite dedicated to hygienic reinvention of the young Latin American nation. Throughout the second half of the nineteenth century, a repeating cycle of cholera and yellow fever epidemics had brought about the emergence of the higienistas, a powerful group of doctors and public health proponents who merged issues of social and cultural progress with the installation of public health infrastructures. With ambitious programs that aimed for heightened standards of household cleanliness, these doctors appealed to a new sense of hygienic citizenship, in which the value of liberalism was joined to strict marital ideals, moral regimes, and utopian visions of public health.[19] To extend "modernity into the flesh" of Argentinean citizens, Ruggiero writes, new laws, regulations, and unprecedented campaigns were devised that relied on intrusive inspection.[20] Additional to the installation of large parks, the construction of a sewage system, and the reorganization of garbage disposal, traditional sanitarians were keen to intervene into private spaces. Instructions for cleanliness in the private home were essential instruments for engineering social change.[21]

The higienistas' extraordinary program of sanitary utopianism has often been attributed to the nation's father of public health, Guillermo Rawson. He founded Argentina's first board of health, the "consejo de higiene pública," in 1852, which was built to resemble European and North American models as a body of independent expertise. Rawson laid the groundwork for an emerging class of a medical elite. Among them was Eduardo Wilde, who had joined forces with Rawson at the university's hygiene department, and who took over the capital's waterworks and sanitation commission in 1880.[22] A strong believer in the promise of sanitary reforms, Wilde continued to improve the capital's cleanliness to bolster its defense against the entry of diseases.[23] The early higienistas shared a belief in miasmatic concepts. Accordingly their popular reforms required a shared responsibility, far exceeding that of physicians, for the hygienic protection of the urban space in its entirety

In 1880 the board of health was replaced by the national Department of Hygiene. Responsibility for the sanitation of Buenos Aires was assigned to the Asistencia Pública in 1883. Inspections, reporting, and control were carried out by its executive arm, the Sanitary Administration, initially led by Drs. Carlos Malbrán and Antonio Gandolfo. With this institutionalization of an ambitious hygienic programme, prevention strategies and prophylaxis were formalized and became enshrined into state doctrine.[24] The Asistencia Pública became responsible for the registration and surveillance of all infectious disease cases.[25] Its executive arm carried out public work projects such as installing running water and a new sewage system, as well as assuming control of hospitals. But it also became responsible for overseeing wet nurses, vaccination programs, transportation of cadavers, and the regulation of prostitution. Public sanitary initiatives, such as encouragement to improve hygienic bathroom practices and install modern toilets, were at the time driven by what Ruggiero describes as a "moral contagion."[26]

The impact of the higienistas started to flourish when "the state committed expanding resources to public health campaigns against epidemics, which in turn provided a model for similar public-private collaboration in the sphere of social hygiene."[27] Wilde had installed a system of hygienic inspections for poor housing with regular visits every two months.[28] He had remained sceptical of the ongoing bacteriological transformation of public health, and much of his enthusiasm for structural sanitary reforms in the sewage system and in private households sprang from his belief in the spontaneous production of epidemics through contaminated conditions such as noxious gases from latrines.[29] Among his most prestigious projects was the establishment of parks and the preservation of green areas in the capital to safeguard its "reservoirs of pure air."[30] Wilde's public health program "deemed the city an artefact and social fabric in which fear of contagion, the morality and living conditions of the urban masses, and concerns about faulty city infrastructure were closely associated."[31] Not only did medicine encourage seeing the urban society as a malleable organism, but doctors also took on the role of social engineers.[32] The historian Myron Echenberg considers the medical elites around Wilde to have been driven by a simplified Darwinism, installing a discourse of hygienic improvement with roots in Neo-Lamarckian ideas of heredity. Improving the nation's health was thus supposed to raise moral, racial, and physical purity.[33]

Wilde's replacement at the Asistencia Pública in 1900 was Carlos Malbrán, a bacteriologist schooled in Europe and a firm believer in germ theory and in the necessity for scientific responses to epidemic threats. After a steep career in the medical institutions of Argentina, his appointment was paradigmatic for the arrival of laboratory science in the public health architecture of Buenos Aires. Malbrán's epidemiological views centered on the microbe, and his policies focused on its containment both at the port and in the urban landscape. His heritage, enshrined today in the nation's bacteriological research institute, the Instituto Dr. Carlos Malbrán, was continued in 1906 by José Penna, then the doyen of Argentinean bacteriology.

Since Penna was promoted into the country's first chair of epidemiology in 1900, he became known as a fierce opponent of the anticontagionist sentiments of Wilde and Rawson. But despite the strong scientific program from Malbrán and Penna, both remained dedicated to the improvement of the city's political, social, and moral constitution by means of prevention and prophylaxis.

José Penna and the Bacteriological Renewal of "Higienismo Argentino"

As he looked back in his 1910 report, Penna placed the establishment of large-scale urban fumigation at the logical end of a history of rigorous surveillance and intrusive disinfection, structured by technological advancements since the late 1880s. As a modern epidemiologist, Penna condemned the intellectual foundations of Wilde's sanitary legacy, but applauded many of Wilde's initiatives that had improved the nation's hygiene. Kohl suggests that by perpetuating the popular legacy of Rawson and Wilde, Penna's brand of epidemiology successfully merged the utopian visions and idealising language of the old sanitarians with the new laboratory science and modern principles of prophylaxis.[34] Accordingly, Penna celebrated some of the established instruments of Rawson and Wilde, such as hygienic education, and particularly emphasized the significance of central civil registries.

For Penna, the registries had been a key advancement to track the success as well as failure of public health measures implemented since the 1880s. The civil registries revealed a general reduction in the city's mortality rate over the 1880s and 1890s. While the population almost doubled between 1888 and 1899 (from 455,167 to 795,323), the general annual mortality only grew from 12.367 to 13.567, which caused a considerable drop from 27.17 percent to 17 percent.[35] Despite this success, infectious diseases remained the cause of death in almost half of the cases throughout that period. While Penna applauded the positive impact of Wilde's reforms, he saw the continuously high death rates due to infectious diseases as an embarrassment for the sanitary state of the capital. "It should not be forgotten," he wrote in 1910, "that these are diseases, that by their very nature, can be reduced or even prevented, some of them completely, with the reasoned application of modes of prophylaxis."[36]

Infectious diseases, Penna explained his epidemiological views further, could be defeated in two ways. First, to establish an individual state of vigorous health, he advised to eat adequate nutritious food, to avoid excess of all kinds, and to distribute vaccines for diseases such as smallpox. But second, the health administration had to develop systematic approaches to destroy disease vectors and exterminate pathogens from their hiding places throughout the city.[37] Isolation and quarantine of patients in pesthouses had only limited effects, as it emerged in 1900 that animals such as rats might be indeed responsible for disseminating diseases-causing agents across the city. Penna concluded therefore, that "[t]he disinfection of places, clothes and objects that the patient might have contaminated, are central measures of prophylaxis."[38]

In turn-of-the-century Argentina, after Wilde's departure, "the microbe" was the new symbolic and practical target of prophylaxis. Miasmatic theories were then dismissed as irrelevant premodern conceptions of disease causality. But early bacteriology in Argentina never fully endorsed an autarchy of the laboratory. Microbes were seen by the country's most eminent epidemiologist as only one of two indispensable factors; the cause of an infectious disease could equally be attributed to the conditions that allow the microbe to flourish. Importantly, this ecology of disease pointed far beyond the constitution of the human body. Where Michael Worboys has modulated the historical narrative of the supposed bacteriological revolution as one that emphasized the relationship of the human "soil" and the bacteriological "seed," Penna's convictions integrated the environment in its entirety as a consideration of a modern, scientific reinvention of traditional hygienic practices in Argentina.[39]

But this appreciation of urban ecology in terms of bacteriological science did—contrary to the perspective of Bruno Latour in *The Pasteurization of France*—not bring about a sharper focus on sanitary intervention.[40] Instead, Penna's pragmatic approach to microbes and their environment was characterized by the fact that knowledge about animal vectors and the implication of soil, merchandise, and furniture in the transmission of infectious diseases remained an object of heated speculation. He admits freely that "the etiological foundations for the advised disinfection procedures are at present not fully known."[41]

To Penna, the vital factor to prevent infectious diseases was *el terreno*, the terrain.[42] This was where the microbe cultivated itself and through which it manifested its properties, and, most importantly, it was the breeding ground of rats. Without a systematic consideration of soil, walls, and ceilings, as well as of fabrics, grains, fruits, and vegetables, Penna considered the fight against infectious disease to be futile. If one were aiming to destroy pathogens, the real aim of public health measures was to remove every possible condition under which the pathogen could survive outside the human body. But given widespread uncertainty about the relationship between microbes and their material environments, Penna conceded by 1910 that the sanitary fantasy of complete destruction of the conditions for microbes to survive would remain a utopia—at least for the foreseeable future.[43] It is important to note here that ambiguity about specific vectors underpinned Penna's epidemiological understanding of the terrain as a broadly defined space in which bacteria was nesting and in which transmission could happen. Uncertainty about etiologies and modes of transmission motivated Penna's improvement and expansion of a disinfection program that would destruct any plausible vector and destroy bacteria as much as insects and rats.

Established in the early 1880s, a steadily growing body of brigades had been responsible for the inspection and control of hygienic standards in food production, but also to improve the general hygienic appearance of the city, both on the streets as well as in private homes and industrial sites. The fight against contagions required, so Armus writes, new forms of social control and implied new practices of socialization in which humanitarian efforts were often met by strict sanitary enforcements.[44] As cleanliness and its restrictive—at times, intrusive—implementation became a cornerstone of the capital's self-perception, sanitary brigades became an iconic representation of the state's efforts to police its hygienic status. Through the politics of the brigades, Rodríguez argues, health came to be seen as the supreme law of the state, and hygiene the appropriate tool for forging necessary social change to prevent epidemics from both arriving and thriving.[45] Infections that appeared in spite of such extensive measure were often attributed to social degeneration, racial impurity, and the subsequent reversal of "Europeanization."

In 1888, the rapidly growing Asistencia Pública invested in the orderly installation of a disinfection brigade, the "Cuerpo de Desinfectadores," to act in rapid response to cases of disease reported by inspectors or suspicions raised by citizens.[46] Upon the initiative of Ramos Meija, disinfection campaigns were thought to bring the benefits of bacteriology to the houses and streets in which cases of infectious disease had been observed. Throughout the 1890s, this procedure became a guarantor of safety and was widely thought reliable. From 1893 onward, restrictive laws and strong legislative capacities gave sanitary forces invasive access into the fabric of Argentina's society. Sanitary police corps secured the city at the turn of the century in its port, its production facilities, its graveyards, and its food facilities.[47] Penna saw the brigades as a decisive factor for the slightly decreasing numbers of infectious disease, and he attributed their success to a seamless system of inspection and subsequent disinfection. He demonstrated his conceptual admiration of the brigade's work in a detailed diagram. A scheme of the standard procedure of disinfection offers a good impression of the meticulous order with which a property was returned to safety after an outbreak was reported (figure 6.1).

After a disease was reported, a series of forms, reports, and surveillance notes were gathered in the office of the sanitary administration, an officer (usually a physician) was sent to the building in which the case of an infectious disease had been reported, to make assessments regarding the general hygienic state of the building, its interior, and its water facilities. Another officer would go to the nearest disinfection station to instruct the crew about the case, so that they could ready one of the prestigious Geneste-Herscher disinfection steamers. The historian Kindon Thomas Meik describes the procedures that followed:

> Disinfection teams arrived wearing a special suit provided by the administration. The suit completely covered the clothing of the employee and the head was to be covered by a cloth hat. Hair and beards were to be trimmed short. Prior to entering the home, the team disinfected the soles of their boots in a bichloride mercury solution. Once in the home, the team removed any contaminated clothing, bedding, and household items that could be placed in a vapour humidifying stove to be cleansed.[48]

The crew would then gather all infected materials, mostly fabrics—such as clothes, leather and linens—and transport them to the disinfection station, where they were exposed to large volumes of steam and then returned in a sterilized state.[49] This protocol foregrounded the urban places of infection and moved the private environment, in which microbes might lurk, to the fore of prevention practices.

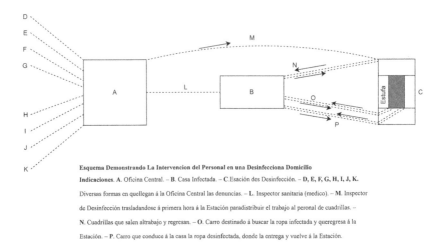

Esquema Demonstrando La Intervencion del Personal en una Desinfecciona Domicilio

Indicaciones. A. Oficina Central. – **B.** Casa Infectada. – **C.**Esación des Desinfección. – **D, E, F, G, H, I, J, K.**

Diversas formas en quellegan á la Oficina Central las denuncias. – **L.** Inspector sanitaria (medico). – **M.** Inspector

de Desinfección trasladandose á primera hora á la Estación paradistribuir el trabajo al peronal de cuadrillas. –

N. Cuadrillas que salen altrabajo y regresan. – **O.** Carro destinado á buscar la ropa infectada y queregresa á la

Estación. – **P.** Carro que conduce á la casa la ropa desinfectada, donde la entrega y vuelve á la Estación.

Figure 6.1
Scheme to present the brigade's interventions in disinfecting a private home. Drawn after the original.
Source: José Penna, *La administración sanitaria y asistencia pública de la ciudad de Buenos Aires* (Buenos Aires: Imp. G. Kraft, 1910), 132.

According to Alvarez, this system of disinfection was publicly considered capable of returning places in which infections had occurred back to a healthy state. It encouraged trust in the health authorities and led rarely to resistance or objections.[50] When the brigade disinfected a house, these practices not only aimed to clean the area of specific pathogens, but the brigades also demonstrated to the public that traditional methods of cleaning and washing, carried out by families in private, were insufficient. Alvarez emphasizes that the health authority did not only aim to ensure the health of the population, but it made sure these actions were impressed on public memory.[51]

The Crisis of Bubonic Plague in 1900

By the end of the nineteenth century, Buenos Aires was considered one of the cleanest cities in Latin America. In 1899 Luis Agote, the city's bacteriologist, assumed Buenos Aires had replaced even the lauded status of London as the most hygienic urban location in the world.[52] But, as Agote admitted later, all the cleanliness did not prevent the eventual arrival of bubonic plague in the city. Echenberg points out that because Buenos Aires was seen by its own elites as one of the healthiest cities in the world in 1900, the "arrival of bubonic plague came as a great shock."[53]

The appearance of this epidemic, first in Rosario and later in Buenos Aires, allowed the medical elites to drastically shift the direction of public health services in the Argentinean capital. The traditional quarantine of both cities, as practiced the year before in Portugal's Porto, seemed impractical. Quarantine posed logistical challenges to a large city and exceeded the government's capacities. Moreover, the ensuing commercial interruption was considered too costly, given the relatively low numbers of infections and lethal cases.[54] Instead, as in a number of other ports in the same period, the outbreak of plague encouraged public health officials to radicalize and extend fumigation practices. Like many other port cities, Buenos Aires had already invested in the installation and development of a sophisticated disinfection station at the port with adequate capacity to accommodate the growing volume of maritime trade.[55] Over the first decade of the twentieth century and in the aftermath of plague outbreaks, these precise routines, developed at ports and disinfection stations, were mobilized and applied across the urban landscape of Buenos Aires to sustain the vision of a disinfected capital, to establish the hazardous presence of an infectious terrain, and to turn the brigade's activities into a coordinated practice of prevention.

Plague in the south of Latin America broke out initially in Asunción, Paraguay's capital, in April 1899. The *Centauro*, a ship under Argentinean flag bringing rice from Bombay via Rotterdam, provoked suspicion. Although it initially disembarked large portions of its cargo in Montevideo, Uruguay, the outbreak occurred later in Asunción, where sanitary measures and surveillance practices were thought mediocre compared to the standards established on the other side of the La Plata River. An anonymous author in the Argentinean medical weekly, *La Semana médica*, articulated a sentiment many medical professionals in Argentina shared at the time: Paraguay simply lacked the necessary sanitary measures that would have prevented the outbreak in the first place.

While the country had been following agreed quarantine regulations of ten-day detention, Paraguay's disinfection capacities were considered inadequate. The disinfection machine available to the port authorities in Asunción, the author reported, was barely large enough to disinfect the personal belongings of a single person, and failed systematically to achieve the sanitary hygiene required to keep plague at bay. Instead, proper fumigation of the *Centauro* with sulphuric gases would probably have kept Asunción and also Argentina safe. This practice was already in place in Brazil's and Argentina's ports.[56]

Immediately after reports of plague in Paraguay reached Buenos Aires, Argentina's medical elite offered its assistance to contain the outbreak. But nonetheless, plague ravaged Asunción from May 1899 to February 1900 and left 114 people dead. As a result of intense trade relations between Asunción and Argentina, cases of bubonic plague were eventually recorded in the industrial Argentinean port city of Rosario. The first patient in Rosario was officially registered on January 18, 1900, but per Agote's report, the epidemic had probably begun in September 1899. In Rosario, the epidemic quickly escalated to approximately 700 cases, of which 248 proved fatal.[57]

Malbrán, a bacteriologist and head of the Argentinean board of health since 1900, was put in charge of the medical commission sent to Rosario, and he declared that poor local conditions had fostered the outbreak. As clinical diagnosis and bacteriological findings were unambiguous, the remaining question for Argentina's higienistas was, how and why plague had found its way into the country, despite the nation's perceived sanitary superiority. Malbrán's preliminary conclusions reinstated the significance of a system of observation and prophylactic intervention in order to prevent the diffusion of the epidemic beyond the port city north of the capital. In addition to reporting suspicious cases and isolation of patients as well as their families, future attention should be given to sufficient disinfection of houses and affected neighborhoods. While it was imperative for boats' cargo to be disinfected thoroughly with sulphuric gases, Malbrán demanded that boats had to be freed of rats by any means available before goods could be unloaded onto Argentinean soil.[58]

By December 1899, plague had probably already arrived in the capital. Although initially registered as a case of severe influenza, Agote declared the fate of the grain dealer "J. M." and identified him as the first appearance of plague in Buenos Aires. It took until January for the epidemic to be officially announced, and a further few months until May for the scourge to have peaked, though it caused a rather low death toll just short of one hundred fatalities. The official response to the epidemic in Buenos Aires was in principle the same as in Rosario: isolation hospitals were set up, workers were sent home, and affected grain depots and industrial areas were closed for extensive disinfection and fumigation. As Echenberg has noted, free public baths were opened and end-of-day garbage collections were started.[59]

But as Agote, Medina, and Penna stressed, the emergence of plague was also indicative of a drastic failure of previously celebrated sanitary measures. Agote and Medina attributed the failed prevention of the epidemic to the etiological complexities of plague. While yellow fever and cholera appeared to have come under control due to disinfection measures—both diseases had largely disappeared from the capital throughout the 1890s—plague eluded the same framework and posed a set of new problems with regard to the introduction of epidemics from abroad. Where methods of maritime sanitation were previously focused on a human carrier, and his or her immediate surroundings, plague shifted attention to the epidemic environment: the ecology of the bacteria responsible for the disease. "The epidemic milieu," Agote and Medina wrote, "is the result of many factors, some large, some rather small, which maintain a perfect balance."[60] The lesson of plague was to approach this milieu in all its complexity, rather than narrowly concentrate on an individual factor or a single cause. The challenge for sanitary prevention was therefore how to disrupt the balance of such an epidemic system in the most effective and efficient way. Moreover, the question emerged, if plague might indeed be an epidemic that was caused by factors within the city and if modes of prevention should move beyond the protection against the disease's import at the harbor.

Here the rat began to enter Argentinean epidemiology. Already by 1899 a few publications in *La Semana médica* had pointed to the growing significance of rats in outbreaks of bubonic plague in India, and in Porto circa 1899. The Pasteurian Paul-Louis Simond had published his research in 1898 on rats as a possible principal vector, a proposition that found increasing recognition in the field but would not be fully accepted until around 1905.[61] Described as an agent of contamination and infection, the rat was usually seen in Buenos Aires as an important concern regarding the import of the disease. M. Netter, a local physician, assumed in 1899 the significance of the rat as a vector but subordinated it to the "contagio directo"—direct human-to-human contagion—yet still considered plague's bacterial cause to be mostly transmitted on the surface of cargo, merchandise, and foodstuff.[62] Diogenes Decoud, also writing in Argentina in 1899, largely agreed with the argument of Simond, and afforded a vague position to the rat as an infectious influence, a possible carrier of the pathogen in the holds of ships.[63]

The rat was mostly seen as a problem for the port authorities. In the immediate aftermath of the outbreak, the rat remained associated with the introduction of plague through the harbor. The regulations established in 1892 to protect Argentinean ports against "exotic diseases" were seen as unreliable as they had failed to integrate the transmission of plague through rats. Agote and Medina stressed in 1901 that a response was needed that would cover both the transmission of plague via rodents, as well as possible transmission through cargo. "Dr. Simond strongly argued that the rat is the main agent for the spread of plague," they wrote, "but recognises the importance of the role played by infested merchandise which would in turn contaminate the rodents, which would thus become the main factors in the expansion of epidemics."[64] Citing the definition of quarantine at the 1897 Venice conference, they argued that although evidence for infested merchandise and rats was lacking, to ignore this hypothesis would leave vast pathways of possible transmission open. "Neither the isolation of the sick, nor the disinfection of the vessel under the prescribed conditions, would have caused the death of the rodents, the principal vehicle of the disease," they added.[65] The conclusion should therefore be a thorough disinfection, supervized by sanitary officers and independent physicians, on every vessel entering from infected countries that could possibly be carrying infested merchandise or infected rats.

Even though the consideration of rats began to suggest that plague might indeed not only be a threat of importation, the city's first reaction was the improvement of the port's disinfection equipment, and in 1900 a fully equipped "Estación Sanitaria" was planned. These considerations, Malbrán remembered in a short essay from 1931, motivated his first actions in the Dirección Nacional de Higiene, as he ordered not only the erection of the sanitation station but abolished effectively all traditional quarantine.[66] The practice of involuntary detention had damaging repercussions for both transnational and national trade, and had failed to protect Argentina from foreign diseases, plague most notably. Instead, effective protection from epidemics from maritime trade should be set on firm scientific grounds. The new station, finished by 1906, was built complete with a library and a "Museo Sanitario" (Sanitary Museum) to train new officers. Until the station was finished, floating fumigation devices were equipped with a Clayton machine and were used to carry out disinfection on arriving and departing ships. They exterminated microbes on surfaces and in the merchandise, but also provided an effective weapon against rats and insects to protect Buenos Aires from further import of infectious diseases.[67]

In the aftermath of the 1900 plague outbreak in Buenos Aires, modern sulphur-based sanitation equipment became quickly standardized in the city's harbor. It was only a matter of time until considerations around the rat as a possible vector of plague in vessels were extended to the many rats populating the streets and burrows of the capital. And as suspicion grew that the city's rats were indeed implicated in the dynamic of plague in Buenos Aires, so grew the conviction that existing urban measures of prevention and intervention were in need of critical upgrades. Over the following years, two decisive changes were made. First, with the acquisition of the Aparato Marot in 1906, the technological advantages of modernized sulphur-based disinfection machines were introduced into the work of the sanitary brigades. Second, this acquisition was accompanied by a conceptual shift, as the brigade's work was less and less seen as a mode of intervention in the aftermath of outbreaks and infections, but would instead become a practice of prevention aimed to establish "general prophylaxis."

The Aparato Marot, the Rat, and the City

The consideration of using machines on the streets of Buenos Aires that were originally developed for sanitation in ports was accompanied, and perhaps encouraged, by a decisive shift in the identification of the cause for plague outbreaks in the capital. Rather than being repeatedly imported by sea trade, plague came to be considered to silently hide within the fabric of the city. Quickly after the 1900 outbreak, the rodent and its burrows were suspected by the higienistas to permanently harbor plague. Agote and Medina discussed already in their 1901 report that plague was perhaps not always imported but could have been sustained by nonhuman vectors within the city.[68] Penna argued in the same year, in a contribution on the etiology of bubonic plague in *La Semana médica*, that, while proof was still lacking, the rat should be considered the most likely source of the epidemic and its fluctuations in the capital.[69] From 1901 onward, the higienistas shifted the government's epidemiological focus from the entry of plague from foreign countries to the conditions under which this and other diseases might nest within the fabric of the urban infrastructure. To address the newfound view, that the city and its rats could harbor disease, the sanitary brigades were reorganized and "the personnel was increased," Penna wrote, "to extend the capacities of home disinfection against the multiplication of transmissible diseases."[70]

For the disinfection brigades that patrolled private homes and warehouses in Buenos Aires after 1888, the steam-based sterilization of goods and fabrics had long been the most important practice. Additionally, the brigades usually used sprayers delivering dichloride solutions to wash walls and floors. Burning sulphur was practiced in a less regulated manner than the majority of other sanitary routines and had been guided by vague and outdated assumptions about the cleansing effects of the fumes, which were never tested for efficacy. While sulphur's principal target had been the disinfection of private homes, it was also speculated to work on "miasma, the effluvia and emanations from fermentation or rotting organic material."[71] As this disinfection practice was carried out poorly, Penna complained in 1910, it was merely effective against pathogens, and acted predominantly "on the frightened imagination of the people."[72] Commonly seen as a practice of cleaning the air from bad odors, Penna had observed a regular fumigation build-up as sulphurous fires filled with "sinister gleams the sad and sombre picture of the capital."[73] But the extensive fumigation of houses affected by cholera or yellow fever using sulphur, chlorine, and nitrous vapors was not based on any sound scientific research and did not, after all, prevent the outbreak of plague. The reason was to be found, said Penna, not only in the poorly executed fumigation, but also in the lack of systematization in the brigade's intervention, which responded only to individual cases and outbreaks instead of providing a large-scale service of preventing disease across the entire city.

Since the 1890s, the sanitary administration had steadily increased its disinfection activities. While in 1889 the service had only disinfected 1,458 houses after infections had occurred, by 1895 this number had increased to 4,542, and by 1905 disinfection teams visited over 7,000 dwellings per year to carry out their work.[74] A mobile disinfection stove was already purchased in 1895 to support disinfection of premises in the north of the country. In 1901 and 1904, both disinfection stations were equipped with a so-called "dehaitre" system that enabled disinfection with formalin and formaldehyde. Once Penna took over the Asistencia Pública in 1906, his steadfast microbiological conviction, coupled with his unique appreciation of the disease's terrain, moved to the center of his unique version of an "Utopía del Higienismo."[75]

Within a few months, the sanitary service saw a major expansion of fumigation equipment designed to mobilize technological appliances and to bring the full scope of disinfection advantages directly to the benefit of the entire city. Three new disinfection stations were opened and equipped with twelve sterilization ovens, and thirty-six new Geneste-Herscher sprayers were bought to maximise the outreach of the sanitary brigades. But most importantly, the sanitary campaigns could from now on also rely on a brand-new fumigation machine. Under Penna and with the support of the new technology, the purpose of the brigades shifted from returning infected premises into a healthy state to the preventive disinfection of the entire city and thus gave way to develop the ambitious goal of a total deratization of Buenos Aires. Not only did the Aparato Marot enable this significant shift of the brigade's purpose, it also reinstated the popular and common practice of sulphurization as a modern technology of disinfection and disinfestation.

Initially, and on advice from Penna, the city bought a pair of model No. 2 machines, and four of the larger model No. 4, two of which were mounted on automobiles, while the others rested on horse carriages. With the acquisition of the Aparato Marot, said Penna, the disinfection brigades were for the first time able to fulfil their envisioned role to the full extent.[76] As the total deratization of premises became suddenly possible, Penna gave the brigades a new purpose as a means of protection and reshaped their practices as a mode of prevention. The machine's design provided quick and mobile interventions everywhere in the city and, as Penna emphasized repeatedly, its mode of disinfection was now reliable as it was set on solid scientific facts. When used correctly, the Sulfurozador was capable of delivering "perfection" in the destruction of everything that could lead to infectious disease: pathogens, insects, and rodents.[77] Pumping this novel gaseous mixture, an electrified version of sulphur dioxide, into houses, restaurants, and warehouses promised rapid and thorough destruction of pathogens, insects, and rodents without harm to organic materials or lasting damage to furniture. There was no reason to apply these outstanding disinfection capacities only to those premises in which diseases had occurred, given the likelihood of rats spreading germs all over the capital.

The Aparato Marot had been built by René Marot in France with the explicit aim to improve the rat-killing capacities of fumigation with sulphur dioxide.[78] These disinfestation capacities were achieved through a specific design. The machine could deliver twenty-five cubic meters of gas per minute. Other than the Clayton machine's reliance on a furnace, it used liquefied gas stored in pressurized containers, which was expanded within a pipe heated by Bunsen burners, which were attached to a pressurized oil heating system. The heating prevented the tubes from freezing due to the drop in temperature during the expansion of the gas. From there, the gas passed through a special device that electrified the gas and thus proposed an extended lethal capacity. The gas was then introduced into the room through a fan that also extracted the air from the fumigated enclosure. The fan as well as the electrifying device were operated by a small combustion engine.[79] This ingenious addition that "ozonofied" the gas gave the machine its colloquial Argentinean name—*Sulfurozador*.

The electrified sulphuric gas had acquired the name Gaz Marot in France. Electrification, so the theory went, reduced the hydrogen in the gas, which might have contributed to fast dissemination as the gas bound itself more easily to existing moisture in the fumigated room. This procedure supposedly doubled the efficacy of the compound against rodents when compared to the Clayton. Experiments conducted in Paris by Wurtz and Bonjean showed that with twenty grams of sulphur used to make conventional sulphur dioxide, rats died after forty-eight minutes of exposure, while the same amount of Gaz Marot proved fatal after only twenty-four minutes. Furthermore, Gaz Marot showed similar lethal qualities to microbes as sulphur dioxide, and was superior in its preservation of by-standing objects, goods, and foodstuffs. The Comité Central des Armateurs de France, the organization of shipowners in France, officially preferred the new Gaz Marot over the Clayton machine after 1905 as it proved efficient against both rats and vermin, as well as harmless to valuable merchandise.[80]

Penna also declared the Marot machine to be advantageous compared to the outdated Clayton machine. The pressure released by the Marot was constant rather than shifting as with the Clayton. The amount of gas required for disinfection of warehouses, tanks, railway stations, docks, and barracks was lower, and the Marot required manageable volumes of liquid sulphuric acid, and the pressurized compound could be easily stored in large quantities. Furthermore, the ozone effect stemming from the electrification of the gas drastically improved its qualities as insecticide and pesticide.[81] Penna referred to this effect as a stirring of the atmosphere, "una agitación de su atmósfera."[82] This prominent feature of the machine was essential for Penna to promise the effective asphyxiation of rodents in every aspect of the urban environment. Where Clayton machines relied on the considerable pressure of closed spaces such as the holds of vessels to achieve similar results, the Marot's technical capacity enabled Penna to reinstate his vision of an entire city without pathogens. Previously a vision that had been a privilege of the old sanitarians' miasmatic views, Penna could reinvent their popular utopia now with the support of modern science and engineering.

Penna further pointed out that the major benefit of fumigation with the Marot was reaching far beyond the problem of rats. Its versatile efficacy could be used against any known pathogen that caused infectious diseases in Buenos Aires at the time:

> The applications of Marot gas are multiple. It will be used for the extermination of fleas, bedbugs, lice, mosquitoes, etc., which can be the propagators of some diseases: for the destruction of insects such as worms, butterflies, moths, etc. . . . Also, it not only possesses these indicated properties, but it is also capable of destroying the microbes of cholera, plague, diphtheria, typhoid, etc.[83]

As the French had already shown, by 1905 the Marot was the best machine for maritime hygiene and port sanitation. But, in the following year, Penna conducted a number of experiments to demonstrate to the world that the machine was equally fitted for disinfecting mills, cereal and grain deposits, forages, railway stations, factories, and, of course, private houses. Penna tested the machine's applicability to a wide range of scenarios, objects, and pathogenic agents. The detailed report from June 15, 1907, sent to the Government of Argentina, is built around the consideration of rats. In the experiment, rodents were placed in cages of varying heights, resembling many of their known habitations across the city. Furthermore, Penna added mosquitoes in nets, cultures of bacteria on agar, infectious substances, and foodstuffs to observe the impact of the Gaz Marot. The test objects were exposed to the gas for half an hour and, in short, Penna could demonstrate the successful destruction of all possible vectors, while foodstuff, clothes, and valuables remained unharmed.[84]

Thus, the Asistencia Pública believed that it had enabled a method for enduring safety against epidemic outbreaks and had provided a new level of hygienic modernity in Latin America, perhaps the world. If applied in the right way, Penna wrote, the machine would allow a mode of prophylaxis to succeed and eventually force all infectious diseases to disappear from the landscape of Buenos Aires.[85] Fueled by the extensive technological capacities of the Sulfurozador, experimentally tested as a recipe for total disinfection, Penna reinvigorated the old hygienic utopianism. But now, the disinfection machine took on the practical as well as the symbolic place of the old sanitary campaigns, enabling the vision of a future without disease, and bringing back the terrain of the city as the central source of concerns. If the new apparatus were used rigorously, if the distribution of fumigating brigades across the city were wide enough, and if a number of additional practices and conventions were applied carefully, Penna argued in 1910, the city would eventually see a future free from epidemics.[86]

In practice, Penna ordered over 10,000 precautionary disinfections of private premises in the first year of his directorship of the Asistencia Pública.[87] He extended orders to use the Aparato Marot to disinfect tram carriages and designed a concerted campaign of deratization, which began at the harbor and should eventually cover the entire capital.[88] By 1908, Penna boasted, 28,101 houses had received treatment from the new machine.[89]

Also in 1908, a paper manual to instruct the military laid out the elements of the new "Higiene Colectiva" (Collective Hygiene) that was driving Penna's vision. The manual gave prominence to individual hygiene and household cleanliness, and discussed how the risks of contamination and infection remained the responsibility of the members of the community. But "[s]trict individual hygiene is therefore to be accompanied by an irreproachable collective hygiene," states the manual, "which must constantly improve the sanitary state of the environment."[90] In the manual, the Aparato Marot was seen as the primary and principled weapon against the most powerful hygienic enemy of the state within its own boundaries: the rat.[91] When applied and used appropriately and extensively, not only would rats disappear, but any possible cause for infection would also eventually vanish. Collective hygiene returned as a shared responsibility for a nation free from disease, but this time it was enabled by the state's modern and capable practice of comprehensive disinfection with the Aparato Marot.

As Pedro Rivero, head of the Disinfection and Sanitation service of Buenos Aires writes in 1911, the transformation of the sanitary brigades under Penna had led to wide-ranging changes. Within five years, Penna had trained and employed ninety new health assistants and ten new foremen. He had raised the budget of the brigades, equipped them with innovative and essential machines and thus prepared the Asistencia Pública to embark on, what Rivero articulated as a courageous vision of the total deratization of Buenos Aires. Building on the legacy of Penna, Rivero was convinced that "[l]os aparatos deraticidas Marot" had fuelled the hope to rid the city of the most important disease-carrying adversary. It had shown how the once utopian idea of deratization had begun to shape the reality of the capital.[92]

Rivero celebrated the introduction of the Aparato Marot as the technological solution for an otherwise unsurmountable challenge. Trapping, poison campaigns, and even specially trained dogs had always failed to penetrate those spaces under floorboards, in basements and between houses, in which rats continued to hide. It was through the fleet of Marot machines that the idea of a full deratization became feasible, as the Marot gas could penetrate every hiding spot above and below the ground. After all, the Aparato Marot allowed the Asistencia Pública to increase its activities far beyond the scope of the traditional disinfection brigades, bringing about an entire new way of precautionary and preventive disinfection. Describing the success of the new campaign so far, Rivero reported that by December 1909 over 1,816 blocks of the city had been disinfected. A map that he published as the service's director in 1911 shows all disinfected blocks painted black. The map and Rivero's celebratory report bear witness to the rapid progress of the new systematic way in which disinfection brigades had covered already one third of the urban terrain of Buenos Aires (figure 6.2).

With the theoretical, experimental, and practical application of the Aparato Marot, the machine encapsulated the key principles of Argentina's hygienic legacy to reformulate them alongside modernized perspectives regarding diseases and their ecologies. The traditional practice of

Figure 6.2
Map of the progress of deratization campaigns in Buenos Aires.
Source: Pedro Rivero, "Saneamiento de la ciudad Buenos Aires. Deratización," *La Semana médica* 18, no.1 (1911): 19. Courtesy of the New York Academy of Medicine Library.

burning sulphur, applied throughout the nineteenth century in the campaigns against bad odors and the stench of latrines, had been adapted in a highly sophisticated practice, mobilizing laboratory science and pragmatic approaches to the niches and rodent vectors of pathogens. With the installation of the Marot, focus shifted from removing smell and stench through sanitary campaigns to the lethal efficacy of a gas in variable environments.

But the Marot also symbolized the urban application of a procedure that was initially developed in the context of international efforts of maritime sanitation. The rigorous implementation of the procedure into the capital's terrain symbolized the extensive political authority of the new public health programs, driven again by a utopia of a disease-free capital. The Aparato Marot was thus not only an instrument of disease prevention, but also a symbol of the reinvigorated state's intrusive capacity to maintain the cleanliness of its citizens' homes. After 1906, as Armus points out, the ubiquitous presence of disinfection brigades under the direction of Penna led to an "army of hygiene and prevention" invading the urban landscape, demonstrating visibly the established links among public health, technology, and progress. In their uniforms and equipped with the powerful new machines, the brigades appeared as a "strange force of occupation" (figure 6.3).[93]

The Aparato Marot served as an instrument of hygienic and cultural calibration for Buenos Aires to keep its streets as clean as its inhabitants' homes, while also sustaining the city's status in what Penna called its "perpetual epidemic state."[94] Finally the Sulfurozador was an essential element of the modernization of the hygienic heritage of the capital. With a prominent, but not exclusive, focus on plague and the rat, it put newfound disease etiologies and theories about their urban hiding grounds on display to the public, while similarly demonstrating the government's masterful control over the city's terrain.[95]

A Theory and Practice of General Prophylaxis

This chapter has shown how the legacy of the Clayton in the shape of the Aparato Marot has reintroduced the particulars of a machine, and its technological as well as its metaphorical capacities, into the history of the hygienic movement of early twentieth-century Argentina. Buenos Aires was certainly not the only place in which fumigation was introduced as a

Figure 6.3
Aparato Marot in operation.
Source: Pedro Rivero, "Saneamiento de la ciudad Buenos Aires. Deratización," *La Semana médica* 18, no.1 (1911): 15. Courtesy of the New York Academy of Medicine Library.

scientific method of urban disinfection. For example, Amador has pointed to the pivotal role that fumigation played in the American intervention in Cuba in the same period. There, the sophisticated practice of the disinfection machines was not only seen as a symbolic demonstration of the American forces' technological superiority, but it also supposedly rejected tired conceptions of racial inferiority and degeneration as causes of diseases. The spectacular fumigation of streets, houses, and sewers in Cuba demonstrated a newfound mastery of the tropical environment.[96] Rio de Janeiro equipped its own sanitary brigades with mobilized Clayton machines, while the same apparatus was used across the United States for comprehensive disinfection of sewage systems.[97] It is, however, difficult to grasp the success of such machines and their associated campaigns. Most cities gave up on their usage, given the uncertainty regarding their efficacy and their high level of intrusiveness.

It was peculiar to Buenos Aires that the apparatus seems to have encapsulated and catalyzed a renewed hygienic enthusiasm for the total disinfection of the city's terrain. We have presented this as a reinvigorated vision of Rawson's 1870s hygienic utopia of a new state free from disease. Penna's technological translation of this vision was realized through the Aparato Marot. A modernized machine, delivering laboratory-developed sulphur-based disinfection, had encouraged a new doctrine of "general prophy-laxis," which indeed continued far into the twentieth century.

In 1912 the "theory and practice" of disinfection became the subject of yet another extensive manual of Argentina's Department of the Interior. Building on a long Argentinean tradition of personal and public hygienic cleanliness, the manual distinguished between two forms of "profilaxia."[98] On the one hand, special disinfection procedures were continued to deal with requirements that emerged from specific diseases. On the other hand, the main area of application for the Gaz Marot was to achieve what the manual called "general prophylaxis." Given its tested and verified qualities, it was the only known substance to be effective against any known infectious agent, all suspicious nonhuman vectors, while at the same time being the least harmful to goods and the interior of premises. The Sulfurozador presented itself as the adequate technological solution, stated the manual's author, to the nation's quest for total hygiene.

The Aparato Marot became a pillar of modernized urban sanitation in Buenos Aires, organized around the question of the rat. Metaphorically, the machine translated a principle of sanitation from the harbor to the streets and homes of Buenos Aires' citizens, establishing the changing acknowl-edgement of epidemic threats lurking within the urban society. But it was the technological novelty and chemical capacity that elevated the machine to embody the principle of general prophylaxis—a principle that attached itself to Rawson's and Wilde's nineteenth-century poetics of hygienic uto-pianism. The Sulfurozador was a spectacular machine that continued to act upon the imagination of Argentina's citizens, but it also enacted a new-found application of bacteriology beyond the laboratory. With the Marot, Penna could successfully modernize and rationalize the traditional sanitary practice of acting everywhere and anywhere against disease. His work con-solidated principles of sanitation, prophylaxis, and the responsible state—principles through which Emilio Coni envisioned later in 1919 the ideal Argentinean city for the twentieth century.[99]

The unique character of sanitary hygiene in Buenos Aires would come to permeate the first half of the twentieth century. Reports, pamphlets, and publications continued the discussion of disinfection practices and contributed to the persistence of a hygienic modernity as a cornerstone of the young republic up to its reinvention under Juan Perón. Disinfection in the name of sanitary improvement, intrusion into people's homes and warehouses, and the imperturbable belief in a protoculture of sanitation continued to structure the self-perception of Argentina's capital. Improvement of the city's sanitary state also continued to be coupled to an imagined improvement in moral and social hygiene. This history of the urban introduction of fumigation machines must be read as a history of the social and technical engineering toward a modern state—a state imagined to be immune to threats harbored in both ports and its citizens' private homes.

7 The Demise of Sulphur

Crowned with the sophisticated invention and application of the Aparato Marot, the history of sulphurization came to a gradual end in the third decade of the twentieth century. The history of Clayton's enterprise appeared to have already ended in 1910. Overwhelmed with debt, the Clayton companies had to declare bankruptcy and their shares were swallowed by their debtors.[1] While the precise cause for the sudden end of his global manufacturing company remains unclear, Clayton's bankruptcy certainly anticipates the demise of sulphur-based disinfection and disinfestation methods. Traces of Clayton's life and work thereafter become scattered. Passenger lists and passport applications from 1907 and 1909 indicate frequent traveling between New York and his branches in Paris and London, but with the end of his company at age 53, his public life as manufacturer of fumigation machines ended.[2] However, his machines would remain in use through WWI, and it was only in the 1920s that a new method began to replace the apparatus to which Clayton had lent his name, and which had defined the sulphuric utopia.

Paradoxically, given the demise of the Clayton Company in 1910, by 1920, sulphur dioxide and its various derivates had become the chemical of choice for maritime sanitation. As a result of the introduction of the Fresch method, sulphur was cheap to obtain across the globe. The resulting gas had been proved harmless to most goods, fabrics, and merchandise exposed to it in the process of fumigation. Due to its significant smell, the gas was also noticeable by the personnel carrying out the disinfection work, and was thus considered safe to use. And with the right kind of machinery, questions of pressure, density, and circulation were refined to guarantee the highest levels of efficiency. Finally, perhaps the most important quality of sulphur-based fumigation was that it was internationally agreed to be destructive to pathogens, as well as to insect vectors and rats, providing an instrument of what came to be established as the trinity of hygienic utopianism: disinfection, disinsectization, and deratization.

So what was it that led to sulphur's demise as a fumigant in the 1920s? It has already been noted that SO_2 had some persistent problems: some derivatives were inflammable, the stench was often considered a lasting nuisance, while the chemical also proved to be highly corrosive to silver and gold. Furthermore, the weight of the gas was repeatedly an issue, as it tended to sink to the ground of fumigated holds, allowing rats to climb to safety, or seek refuge in rat-permeable infrastructures (called "rat harborages" in relevant literature) where the gas was unable to penetrate. All this made the efficacy of SO_2 as a fumigant a much-contested subject. This was especially so as, in the aftermath of World War I, the priorities of fumigation and, more broadly, maritime sanitation appeared to shift from disinfection to disinfestation. This proved to be a critical moment in the history of fumigation and the hygienic utopias that had until then propelled the entanglement of experimental systems, maritime regulation, imperial competition, and trade interests around sulphur. For as we have examined at length so far, the success of sulphuric fumigation was predicated on the way in which it appeared to provide an optimal solution to a combined hygienic goal: the destruction of germs and their vectors. This chapter will examine how the gradual abandonment of germicide as a goal of maritime sanitation and the refocusing of fumigation singularly on deratization led to new entanglements between scientific, engineering, economic, and political goals and processes. These found material grounding in the shift from sulphur to cyanide as the fumigant product of choice, initially in the United States and, by the late 1920s, across the globe.

The Rat Refocused

As historians of sanitation in the United States have noted, the image of the rat developed in North America during the first half of the twentieth century was heavily racialized and dependent upon prebacteriological investments of filth and decay.[3] The opening paragraph of the Farmers' Bulletin 896, published by the US Department of Agriculture in 1917, gives a clear image of the entanglement of these notions with the specter of plague:

> The rat is the worst animal pest in the world. From its home among filth it visits
> dwellings and storerooms to pollute and destroy human food. It carries bubonic
> plague and many other diseases fatal to man and has been responsible for more
> untimely deaths among human beings than all the wars of history.[4]

Such narratives weaved an image of rat's relation to humanity as one underlined by human culpability. Not only was humanity seen as hitherto unable
to battle the rat in an efficient manner, it was also accused of actively aiding
its encroachment into humankind's realm: "For centuries the world has
been fighting rats without organization and at the same time has been feed-
ing them and building for them fortresses for concealment."[5]

Writing in 1913, R. H. Creel, the Passed Assistant Surgeon of the US Public Health Service, opened his review of the rat as "a sanitary menace and an
economic burden" by stating that "of all the parasites that have their being
in and around the habitation of man the rat has less to justify its existence
than any other."[6] Seen as "devoid of any redeeming traits," the rat was condemned as "a greater pest" even to the fly "because of its depredations and
its possibilities for harm in the transmission and perpetuation of bubonic
plague in a community."[7] In accordance to this narrative, sylvatic sources of
plague, such as ground squirrels or marmots, posed no significant threat to
human health in themselves but only insofar as they were "the source of a
continued reintroduction of the disease, among the neighboring rat population."[8] The doctrine developed, not only in the United States but across
the globe, was summarized in the slogan "No rats, No plague":

> PLAGUE, RAT, FLEA, MAN, PLAGUE, RAT, FLEA, AND MAN AGAIN, so the cycle
> runs unless interrupted. KILL THE RAT AND THE FLEA and there will be no
> plague.[9]

If, as Creel noted, "a ratless country seems almost Utopian"; a "crusade
against the rat" could be a means of "safeguarding the country from any
possible plague invasion."[10] "IT IS NOT TOO LATE TO BEGIN THE WAR-
FARE NOW"—declared the famous dermatologist, leprosy expert, and veteran president of the American Medical Association, Isadore Dyer, in a
1912 bulletin issued by the Medical Plague Conference Committee of New
Orleans for public instruction and wide publication across the medical and
lay press: "NO RATS, NO FLEAS; NO FLEAS, NO PLAGUE! MAKE THAT THE
SLOGAN!."[11]

Growing concerns about rats and maritime trade built up over the first
decade of the twentieth century. They were particularly fueled by studies

showing the rat's ability to jump, swim, and tightrope-walk, as well as by ideas that plague occurred in rats in a chronic form, and observations regarding the propensity of plague-sick rats in particular to remain in bales or other cargo in the process of loading and unloading.[12] A South African report from 1909 stressed:

> A rat sick from Plague may enter and die in a skeleton crate or in a bale of forage, and the carcasses may be carried long distances by sea of rail. On arrival of the bale or crate at its destination local rats investigate its contents, perhaps devouring the carcass of the dead rat or becoming inoculated by fleas which have left the carcass and thus become infected.[13]

Indeed by 1910 the connectedness between ship and rat had assumed almost mythic proportions. Scientific works attempted to link rats and humans in a way that did not simply reflect what, following medical anthropologists Hannah Brown and Ann Kelly, we may call the "material proximities" of their interspecies existence, but also seemed to foster a world-historical connection between the two species.[14] Referring to the journey of Noah's ark, William Hobdy's contribution to the voluminous work *The Rat and Its Relation to the Public Health*, published by the US Public Health and Marine-Hospital Service in 1910, began thus, inextricably linking humanity's and rat's destiny:

> Since men first went down to the sea in ships the rat's voyage-making tendencies have been known, and their fecundity is as well established as their fondness for travel. The record does not state that there were more than a pair on the ark at the beginning of her voyage, but the chances are better than even that her skipper began that voyage with more rats than his manifest showed; but whether he did or not, we can be sure he had more at the end of the voyage than at the beginning. Whether or not succeeding generations inherited from their forbears on the ark this well-known wanderlust is undetermined, but it is a fact that the intimacy and companionship established and begun then have been persistently maintained by the rat ever since. His travels have been coextensive with man's, until to-day there is not a port on earth where the rat is not present.

From Sulphur to Hydrocyanic Acid

William J. Simpson's 1905 influential *Treatise on Plague* brought the rat systematically and forcefully into the center of the modes of dissemination of plague. There, he considered the Clayton apparatus to be a "weapon of utmost value" to the destruction of rats and fleas in the holds and cargo of infected ships.[15] But, as a report from 1910 indicates, Simpson must have also considered the adaption of hydrocyanic acid gas in British India for the means of plague control in the final years of the nineteenth century.[16] During his investigations in British India, he ordered experiments to be undertaken to determine what density was required to use hydrocyanic gas to exterminate bedbugs. As William David Henderson Stevenson explains in his 1910 report on the *Killing of Rats and Rat Fleas by Hydrocyanic Gas*, the bugs were found regularly in premises in which plague had occurred. Simpson briefly considered the bedbug to be one possible vector of plague and thus encouraged its experimental destruction.[17] But due to the relative decline of plague in India and the departure of Simpson from the subcontinent in 1898, experiments had been discontinued and widely forgotten. In May 1909, upon the suggestion of a Captain W. Glen Liston, Stevenson began new experiments to determine the capacity of the gas to kill fleas and rats. The report stands as the first systematic investigation into the capacities of the hydrocyanic acid gas to kill rats, as well as rat fleas, in enclosed compartments.

As observed by Stevenson, and established through work of the Government Entomologist at the Cape of Good Hope, however, the gas did not seem to affect plague bacteria. Nonetheless, Stevenson carried out two series of tests with the gas, which was put into action to disinfect the clothing of travelers on railways, to destroy fleas and rats, and to disinfect plague-stricken houses.[18] For the first experiment, Stevenson mixed potassium cyanide with sulphuric acid and water. He heated the solution over a Bunsen flame and used tubes to direct the undiluted gas directly into a room in which the clothing was stacked. Through a second pipe, air was introduced into the mixture. The room was sealed, and he could quickly establish that a sufficient penetration of clothes was achieved when they were hung up, and that a short period of exposure—sometimes just five minutes—was enough to show all fleas to be killed. The second series of tests was conducted on plague houses, which were sealed as much as possible. The houses had floors of stumped earth, with rat burrows meandering through the floor's subsurface and leading to small openings on the ground. Both rats and fleas were kept in the burrows, the whole setup being designed to resemble the setup of Captain Gloster, which had been originally designed for experiments with the Clayton apparatus. Pumping the gas into the house yielded satisfying results. But it also created noxious vapors around the building during and after the experiment. Stevenson considered the use of hydrocyanic acid as capable for killing fleas and rats in such environments, but did not recommend its application in real settlements, as possible effects on bystanders and neighbors were highly probable.[19]

Historically, hydrocyanic gas had found its way into pest control by American entomologists, who were familiar with the substance from botanical and agricultural deployment. The gas had been used in agriculture in the United States since 1886. In California, Coquillet had originally devised a method to apply cyanide, whose properties were well known to gold and silver washers, to the task of fumigating citric trees so as to rid them of "Cottony Cushion Scale."[20] After all other existing gases and common substances like sulphuric gases had been shown to have adverse side-effects or no effect at all, Coquillet devised a method of dissolving cyanide in sulphuric acid with water to effectively fumigate trees with the resulting hydrocyanic acid gas. Soon after, the gas was picked up by South African government entomologists who conducted experimental fumigations of prisons in order to rid them of *Acanthia lectularia*, the bedbug. Although the particular gas was considered too dangerous for house disinfection, it was used in 1900—by request of the Colonial Office—to fumigate the Worcester jail. In 1901, Tokai and Kimberley, two other prisons in South Africa, were treated in the same manner. Although these trials were considered successful, the amount of potassium cyanide required proved to be expensive both to purchase and to use.[21] Nonetheless, the colonial authorities proceeded to experiment with the cyanide-based fumigation on rail carriages to destroy vermin of all kinds.

In the field of botanical application, some experience had been gathered on the correct proportions and the particulars of using and controlling the dangerous substance. For Stevenson, the relationship between potassium cyanide, sulphuric acid, and water should follow the already established "1-2-4 formula," which left enough water to dissolve potassium cyanide fully and was also seen as a proportion in which the temperature of the solution would yield to the most desirable effects.[22] For the application in boats and buildings, Stevenson remarked for his experiments in Calcutta, the preparation was largely comparable to the established routines of fumigating with sulphuric acid. But particular attention was given to the containers in which the chemicals were to be mixed. The carrier pot needed to be made of glazed earthenware or china washhand vessels, as other surfaces tended to crackle and tin cans were seen to dissolve. Once the pot was placed in the middle of the fumigated compartment, the water was mixed with the sulphuric acid. Adding potassium cyanide to set the mixture off was a risky procedure: "at once drop in the potassium cyanide at arm's length, and run out of the room."[23] Some suggested to wrap acidic salt in paper so as to extend the time it took for the activation of the solution. Others built string and pulley constructions with which the salt could be lowered into the pot from outside the fumigated room. The gas disseminated quickly and was recognizable up to one hundred feet from buildings, utmost attention therefore being needed to restricting humans from entering the zone of danger. Following the gassing, the duration before opening of the compartments depended on a variety of factors.[24]

Insects, rats, fleas, eggs of insects, and lower animals were immediately killed or, as Stevenson called it, "devitalized."[25] As previous application in botanics had shown, the effects on plants and grains, as much as on food and water, was harmless, and later ingestion of all of the above was inconspicuous. But as experiments with various cultures of streptococci, typhus, coli, and plague bacilli showed, the gas had almost no effect on bacteria. Almost all cultures seemed unaffected and continued to grow on agar after extensive exposure to the gas. Nonetheless, according to Stevenson, this did not rule out the use of hydrocyanic acid for the prevention of plague. It was an excellent, cheap, and quick method for disinfecting clothing and fabrics, ridding them of fleas, and also for exterminating rats in rooms, holds, and compartments. Hydrocyanic acid gas was seen as a superior pulicide and as a widely effective instrument of deratization. If its application was followed by the book and supervized by experienced operators, its otherwise immense danger to humans could be effectively controlled.

In terms of the practices and procedures, the utilization of hydrocyanic acid gas in maritime sanitation appeared to be highly compatible with the much older, but still practiced, method of burning sulphur directly in the holds of ships. Following the 1903 sanitary conferences in Brussels and Paris, US port authorities had continued to utilize sulphur in its simplest form, or so-called SO_2 "pot method," for fumigating vessels.[26] Mostly used in ports not equipped with a Clayton machine, or for ad-hoc fumigations beyond the capacities of quarantine stations, the practice had been a widely used alternative to the mechanized procedures of the Clayton. In the words of Hodby, the Passed Assistant Surgeon at Angel Island Quarantine, "For an

empty vessel nothing is so satisfactory as the pot and pan method of generating the gas. It has the following advantages; is more rapid than any other,
is cheaper, is more effective, and is equally applicable to the largest and the
smallest vessel afloat."[27] Hobdy calculated that the best method involved
the use of a six-inch-deep pot with a wider mouth than base (at a 24:16 inch
ratio), which would then put in a tub of galvanized iron (diameter: thirty
inches) containing a small amount of water. Pots should be filled with sulphur shaped in such a way that the top was hollowed into a small crater,
where four to six ounces of alcohol was added. The latter was ignited with
the help of a match and the compartment closed allowing the gas to seep
through.[28] Circulars of fumigation and corresponding instructions to companies operating vessels sailing into US ports from plague-suspect origins,
such as Havana, specified the amount of sulphur to be used in accordance
to the pot method, as well as timing and other practical requirements to
great detail.[29] In reality, however, the use of this method was found time
and again to leave much to be desired. This was especially so when it came
to the destruction of rats. In the reporting year July 1911 to June 1912,
the New Orleans port authorities inspected 110 vessels derived from suspected or infected "plague ports," out of which thirty-one were found to be
"comparatively free from rats."[30] The authorities applied "fumigation with
sulphur," resulting in the death of 537 rats. Most interestingly, the report of
the New Orleans Quarantine Station affords an image of the less-than-neat
reality of fumigation:

> There were several instances of vessels discharging large numbers of rat eaten
> sacks, and very few or no dead rats having been found; and of rats taking to the
> boats when vessels were being fumigated. On several Nitrate boats, it was stated,
> that the rats not being able to endure the fumes from the cargo, took refuge in the
> engine room and cola bunkers, and in that way escaped death by fumigation.[31]

Another contemporary report, this time from the port of New York, evinced
a similar ambivalence in its account of the fumigation of 127 ships over the
previous twelve-month period aimed at the destruction of rats and the protection of the port against plague.[32] As a result of this operation, 1,934 rats
were recovered, out of which an astonishing but no less unusual 1,224 were
examined in the lab, with seven cases "showing suspicious forms," leading
however to no positives in inoculated guinea pigs.[33] If this all sounds pretty
ordered and neat, a different picture arises from the description of the fumigation process itself. The report confirmed that the number of rats recovered

far from corresponded to the total number of rats destroyed as a result of the operation: "In many cases all the rats killed are not found. Eleven of the ships subjected to fumigation were fumigated at quarantine on their way to sea and the sulphur gases being not sufficiently clear from the hold to allow a search for dead rats, the vessels were permitted to depart."[34]

The practical problems facing fumigation, as evident from the archives of the US Public Health and Marine-Hospital Service, were epitomized by the fact that vessels did not possess structures that would allow for the fumigation of their holds. The New York & Porto Rico Steamship Company was thus urged to take urgent measures of retrofitting its boats so that each and every part of them would be "either capable of being opened up for examination or to allow the free circulation of sulphur fumes."[35] This required port authorities to train shipping companies to read their vessels in relation to rats:

> The ceilings in the holds for instance is an excellent rat harbor, the planking is usually spiked down or very securely fastened so that it is very difficult to remove any of them. The pipe casings in the holds also are securely nailed and are not only rat harbors, but serve as runways for rats from one part of the ship to the other. On your vessels it frequently requires three or four hours work by the carpenter to open up these places preparatory to fumigation, and I would suggest that panels be placed in the ceilings, each panel with sufficient ring-bolts, so that say a quarter of the entire area could be removed easily.[36]

A more elusive problem concerned operating on vessels that had been subjected to different methods of fumigation at their origin and in their intermediate stations. For example, the British SS Bessie Dollar arrived on July 26, 1912, at the quarantine station of Angel Island. A portion of the vessel had been fumigated in Manila, the Philippines, though it was not detailed which parts of the ship this included. Later, when the vessel stopped in drydock in Hong Kong, the harbor master did not allow fumigation; as a result she was subjected to fumigation with "sulphur gas" upon arrival at Angel Island. As, however, the vessel "was fumigated while loaded it was not practicable to search her for rats," the result of the process being simply declared to be "unknown."[37]

Steamship companies in the United States were keen to avoid fumigation. Despite evidence to the contrary, they continued to stress that this was impractical insofar as their vessels were rarely if ever empty of cargo, alleging that fumigation would damage the latter. A series of responses by steamship companies to the US Public Health and Marine-Hospital Service from the summer of 1909 indicated that rat-catching by professional teams and the use of poisons, as used by themselves on a regular basis, were a sufficient alternative.[38] Objections raised against fumigation included opinions regarding the rat's ability to escape the fumes to other compartments, as well as concerns that rats dying as a result of fumigation did so in places from which they could not be retrieved.[39]

Another tactic by shipowners and captains involved invoking the presence of cats on board as sufficient protection against rats. In a rather morbid display, Surgeon G. M. Corput responded to such claims by publishing photographs of the outcome of fumigating the British ship *Ethelhilda*, which arrived in the quarantine station of New Orleans on March 18, 1914. The captain of the boat had boasted that his vessel's cabin held no rats as a result of "the presence of an exceptionally good cat." Forcing the boat (including its cabin) to be fumigated, Corput displayed a morbid photo of the resulting twenty-four dead rats alongside the carcass of the unfortunate if now famous cat (figure 7.1).[40]

CAT AND RATS FROM CABIN OF S. S. ETHELHILDA.

Figure 7.1
Photograph of the display at the Upper Quarantine Station in New Orleans, outlining the uselessness of cats as protection against rats.
Source: Anonymous, "Ship rats and plague," *Public Health Reports (1896–1970)* 29, no. 16 (April 17, 1914): unnumbered plate following p. 928.

This does not however mean that all companies followed such evasive tactics. The Boulton Bliss & Dallett Company, for example, reported that its steamers were "fumigated regularly every four weeks. Those which call at San Juan are fumigated at Puerto Rico, while those that do not call at that port, are fumigated at New York."[41]

Indeed, Puerto Rico proved to be an important locus for US developments in maritime fumigation. The 1912 plague epidemic on the island, where an overall fifty-five human cases occurred, led to concerted efforts to halt not only the spread of plague from San Juan to the US mainland, but also to prevent the disease from moving inland in the island itself, and thus to form permanent reservoirs, as had recently been the case in California.[42] Resisting calls to burn down the infected Puerta de Tierra barrio (on account of this resulting only in the rats escaping and contaminating other areas of the city), among the measures taken was fumigation with a new agent, aimed at overcoming the lengthy exposure needed when using SO_2: "cyanide gas" as hydrocyanic acid was commonly known. This was used "either on lighters or in a galvanized iron shed for this purpose."[43] Given the success of the operation in Puerto Rico, cyanide fumigation gradually moved from being considered a method too dangerous for general employment to a serious alternative to sulphurization. Key to this transformation was the 1914 plague outbreak in New Orleans. After outbreaks in Puerto Rico and Cuba in 1912, authorities in New Orleans had already ordered a preemptive rat-catching and trapping campaign and had within days found rodents indeed infected with plague. But despite steadily increased antirat activities, human cases appeared in June 1914, and a modest outbreak escalated over the following months. The outbreak triggered an unprecedented antirat campaign, with over 380 men inspecting 6,500 railcars and 4,200 buildings, fully fumigating 101 ships, trapping 20,000 rodents, laying nearly 300,000 poison baits, and discovering seventeen infected rats.[44]

Norman Roberts described two cyanide fumigation methods employed in New Orleans in 1914, both of which involved the decomposition of potassium or sodium cyanide by sulphuric acid ($KNC + H_2SO_4 = HCN + KHSO_4$) at a ratio of "1 ounce of KCN . . . for each 100 cubic feet with 1⅔ ounces (about 1 fluid ounce) of sulphuric acid, and 3 fluid ounces of water."[45] The first method, known as the "crock" or "solid cyanide" method, was used for smaller spaces, such as living quarters or storage rooms. It involved the dilution of sulphuric acid in water, whereupon the solid cyanide was dropped. This was applied to a sealed compartment, with the cyanide being dropped through an aperture that was then quickly sealed. Though practical in small compartments, this method was not employed in the holds of ships as, first, that would require an extended presence of the cyanide operators in the high-risk area, and, second, it would allow the gas to escape from the top of the hold. The second method, applied to holds, was known as the "barrel" or "liquid cyanide" method. It involved the dilution of cyanide with water in a barrel positioned at the lowest possible point in the hold, with the help of a rubber hose. After pouring the acid from a safe distance (the operators being positioned on the deck of the vessel), a strong sodium carbonate solution was added, its purpose being to expel "part of the dissolved hydrocyanic acid from the waste and reduc[e] the remaining acidity, thus economising on the expensive cyanide and rendering the waste less poisonous, corrosive, and troublesome."[46]

Again, hydrocyanic acid was praised for its ability, upon being mixed with air at 0.4 percent, to exterminate rodents and insects, in particular fleas: "for use against plague it has the great advantage that it penetrates most articles of cargo without damaging them, killing the vermin, no matter how deeply hidden."[47] The main worry as regards the application of this method arose from the fact that, as the particular gas is lighter than air, it risked escaping from the upper parts of the fumigated spaces, while at the same time being too low in concentration in the lower parts for it to kill the target organisms. Proposed solutions to this problem included the operation of a fan that would obstruct the rise of the gas.

Deemed to be superior to sulphur dioxide, insofar as the process involved no fire while its gas evinced better penetration without any damage to the cargo, and to carbon monoxide, insofar as it killed fleas (and also required no fire), in practice the employment of hydrocyanic gas was nonetheless initially limited. To give one example in the context of anti-plague work in New Orleans in the week ending November 28, 1914, only two vessels were fumigated with the gas, in comparison to twenty-four fumigated with sulphur and ten with carbonic monoxide. As the instructions by the US Surgeon General for the use of this fumigation method indicate, the likely reason was the high danger of the gas to humans, which required highly skilled labor and a disproportional degree of attentiveness by the operating agents by comparison to the other methods.[48]

Also important was the high price of potassium cyanide and sulphuric acid, especially in purified form. To try and alleviate this obstacle, experiments on a reduced percentage of the gas were carried out by Creel on rats trapped in wire cages and placed in different positions in ship storerooms and holds filled with personal objects and cargo respectively. In sum, the results of the nineteen experiments conducted showed that it was not necessary to use ten ounces of potassium cyanide per 1,000 cubic feet of space but only five, at a cost of 8.50 USD to 12.50 USD for a space of 100,000 cubic feet, by comparison to a cost of sulphur fumigation for the same space being 13.00 USD.

Among other things confirmed by Creel's experiments, was the higher ability of the gas to penetrate cargo by comparison to SO_2: "Sulphur dioxide, while fairly effective, is not very penetrating. It diffuses very poorly, and in actual practice it has seemed that air pockets in articles of cargo or between packages will afford to rats a sufficient protection against the effect of the sulphur fumes."[49] The "dumping fixture" method of hydrogen cyanide (HCN), involving the lowering of a tin-plate canister containing half-ounce HCN discoids down the holds, and the opening of the lid with the help of a wire or rope, contrasted with the expensive and bulky apparatuses required in SO_2 and carbon monoxide fumigation methods alike. Hydrocyanic acid gas was praised as both the most penetrating and the most toxic among alternative fumigation gas, as well as for being "easily and quickly generated, requir[ing] very little apparatus, [and not being] destructive to inanimate objects."[50] Little attention was paid by comparison to the fact that "attempts to destroy bacteria with this fumigant were unsuccessful."[51]

Clearly in support of HCN over other methods, Creel thus proceeded to tackle the last remaining obstacle to the adoption of the gas by stating that "The element of danger to human life is more or less speculative, and will vary according to the care exercised in performing the fumigation."[52] In particular, he held that the gas posed no danger to the people operating the fumigation process, as long as they made a swift exit from the compartment, and after the fumigated space's apertures had been reopened for thirty minutes.[53] As was the case many years earlier, when the risk posed by the Clayton machine was discussed in Dunkirk (see chapter 3), in support Creel's health and safety reassurances was furnished a colorful anecdote involving a drunk sailor:

An accident on board a ship at New Orleans throws further light on this subject. This occurred during the fumigation of a super-structure on board. The room had a capacity of approximately 1,000 cubic feet. The cyanide was placed in the acid solution and the doorway sealed. A drunken sailor coming aboard threw open the door and entered. How long the man was exposed is uncertain. The exposure was not more than 15 minutes, and possibly only 5 minute in duration. When discovered he was lying on the floor beside the cyanide container. It was likewise uncertain whether he had been overcome by the gas or had lain down in a drunken stupor. When removed from the room he was resuscitated.[54]

Emboldened by these results, but also uncertain about the application of experimental findings in real-life maritime fumigation, "however painstaking the attempt may be to simulate the natural," Creel proceeded to conduct what he called "a true test of efficiency . . . applied to the procedure as carried out in routine practice."[55] What in his opinion made this feasible were the particular circumstances in postepidemic New Orleans at the time: "first, the fumigation of a large number of vessels at the port of New Orleans and at the Service quarantine station at the mouth of the Mississippi River; second, the availability of a large and experienced force of trappers at New Orleans."[56] The latter were the result of concerted measures by the Public Health Service against plague in the city. Creel's plan was to take advantage of the intensive, systematic, and closely supervized rat trapping operations on vessels following fumigation, so as to ascertain the efficiency of different fumigation methods employed. The record involved 214 vessels out of which sixty-two were treated with SO_2 and 182 with cyanide. The results showed that the former method killed 77 percent of ascertained rats, while the latter killed 95 percent. The disparity between the two methods was accounted by their relative success in the case of "superstructure" (storerooms, crew quarters, cabins, etc.), where sulphur's efficiency was reduced to 55 percent, while cyanide's remained as high as 94 percent (by contrast the difference in empty holds was 96 percent to 99 percent, and 64 percent to 80 percent in loaded holds).

Maritime Fumigation after WWI

In the meantime and with the start of World War I, the chemical compounds used in fumigation were closely inspected for possible application to warfare. But both sulphuric acid gas and hydrocyanic gas were in principle incomparable with the practical if no less brutal requirements of chemical warfare, as both gases evaporated too quickly in open space.[57] A variant of hydrocyanic acid gas, concerns over hydrogen cyanide were a constant theme for most of the war, due to its high lethality and the known dangers surrounding its application in fumigation.[58] It was well known to experts on both sides of the war that the gas acted directly on the nervous system and that exposure to high levels of the chemical led to almost instant death. Stories abounded about the Germans developing cyanide bombs, and rumors spread that abandoned trenches were flooded with the gas, "so that the advancing Allies would die like rats."[59] But it is widely accepted that no deaths from cyanide gas were reported in World War I, as the compound was too light to ever be effectively used on the ground. When the French tried to build shells with the gas, they struggled with its low weight and failed to provide an operational weapon. By contrast, what appeared to be a chemical warfare method transferrable to deratization was the use of asphyxiating gases like chlorium and chloropicrin. In his doctoral thesis on the use of war gases for deratization, Étienne Grégoire recounted that chlorium was tested on rat burrows by the inspector general of hygiene in France, Dr. Bordas, in Nanterre, and by Dr. Tanon in the lab. An exposure of five seconds to the gas resulted in the death of experimental rats eight to ten hours later, and also to the immediate death of fleas. In turn, chloropicrin, CCl_3NO_2 (also known as nitrochloroform or trichlorononitromethane), was a substance discovered in 1848 by the Scottish chemist John Stenhouse whose lachrymatory properties made it one of the first chemical weapons to be used in the context of WWI. Whereas in small quantities the substance operates like a tear gas, in large quantities it has suffocating properties: "It was used chiefly because of its peculiar property of causing vomiting when inhaled, thus inducing the soldier to remove his mask and expose himself to the action of gases which penetrated the mask less readily."[60] It was for the latter that in 1917 two Italian scientists, Piutti and Bernadini, first tested the gas on vessel-borne rats.[61] At the same time, in wartime France, Gabriel Bertrand of the Institut Pasteur conducted his own experiments, which would be communicated—with a delay due to war-related secrecy—on February 9, 1920, to the French Academy of Sciences.[62] The follow-up experiments evinced the ability of the gas to kill weevils and rats, but even more its efficacy in the extermination of fleas. Following the condemnation of hydrocyanic acid by the Conseil supérieur d'hygiène publique in 1922, as a result of the experiments of Bonjean in Marseilles, chloropicrin rose in popularity in France.[63] A decade later, the US Principal Chemist, R. C. Roark, would summarize the advantages and disadvantages of the gas as follows:

> The advantages of chloropicrin as a fumigant are: High toxicity to many species of insects and rats; fungicidal and bactericidal properties; complete freedom from fire and explosion hazards; low solubility in water; ability to penetrate bulk commodities; non-reactivity with metals, fabrics, and colors under fumigating conditions; and a pronounced odor and lachrymatory effect which usually effectively warn of its presence. Its disadvantages are: Slowness in action, as compared to hydrogen cyanide; tendency to act detrimentally on living plants and seeds; difficulty in removing its odor from fumigated commodities and spaces; and nauseating effect upon the operator.[64]

However, at the same time as a plethora of sulphur-based apparatuses and processes were coming to replace the Clayton, HCN continued to lead the revolutionization of fumigation.[65] The true real-life test for HCN as a deratization fumigant in the United States came in 1921, when San Juan was struck by another plague outbreak, believed to have arrived from the Canary Islands.[66] This came to reinforce trust in HCN, whose use was employed in the harbor on all incoming Spanish vessels. This centrally involved fumigation of cargo in open lighters, where cargo was covered in tarpaulin in a tent-like manner, introducing cyanide gas underneath, or in closed lighters, where the gas was introduced after their openings had been sealed.[67]

Both the San Juan incident and the outbreak of plague in Barcelona, Spain, in 1922, led American authorities to adopt more diligent measures. In New York, this involved the institution of a strict procedure of fumigation, whereupon being visited by the chief of the fumigation division, it was decided that "the vessel can be fumigated without removing any cargo."[68] Fumigation of the entire vessel began at 9 a.m. This would involve an exposure to HCN for four hours with the help of an "aerothrust" blower, aimed at keeping the gas equally distributed.[69] For this process to be successful, all "enclosed space" (double walls, etc.) was ordered to be opened in preparation. Additionally, the port authority expected guards on watch to report any remaining live rats observed following the discharge of cargo; if such were seen, the vessel was subjected to refumigation. The same process of "fractional fumigation" was to be followed whenever the fumigating officer observed that "the cargo is such that the gas will not penetrate thoroughly."[70] The regulations also provided a solution for cases where the cargo of a given ship completely filled the holds, making fumigation difficult. Sufficient cargo being discharged and fumigated separately on lighters, the fumigation of the remaining cargo could be performed in the holds. In all cases, the ship was fumigated with cargo in situ. Following the unloading of the cargo on the harbor, the ship would be again subjected to fumigation on empty holds.

However, what proved more difficult was to ascertain the real number of rats killed as a result of these procedures. The Staten Island Quarantine Station Surgeon, Grubbs, brought this to the attention of the US Surgeon General already in early 1922. He stressed that the number of rats sent to his lab, resulting from cyanide fumigation of vessels, "is not as large as it apparently should be if we are doing good fumigation and if we are not over-fumigating certain vessels."[71] The problem, according to Grubbs, lay with the fact that stimulating interest among the crew in the recovery of killed rats proved very difficult. On the one hand, he surmised, this was due to the fact that once fumigation was complete, "the crew is in a hurry either to get to the next ship or to get back to the station and be checked out."[72] Efforts to elicit rivalry among the crew in rat recovery proved equally impotent.[73] However, another important factor, perhaps understressed in Grubbs's report, related to the fact that especially when it came to vessels from Puerto Rico or the Pacific coast of South America, frequent fumigation meant that no rats were resulting from such processes, something that "decrease[d] interest" in their recovery.

Following the conclusion of World War I, the continued integration of hydrocyanic acid gas, and the experimental testing of chlorocyanic gas, as well as fumigation with bromide sealed the fate of sulphur. With the disjuncture of disinfection and disinfestation in the practice of fumigation, efficiency in terms of the rapid destruction of pests and vermin became more important than the capacity to kill pathogens. With HCN at the forefront, the focus shifted to insects and rodents; bacteria, by contrast, were considered to be less of a concern, or could be taken care of through washing with carbolic acid. By contrast, combining the expertise of agricultural fumigation and maritime deratization, the application of cyanide compounds in the fight against rats and other rodents made a landfall in the United States. Large-scale campaigns against ground squirrels attempted to exterminate rodents from entire rural landscapes. The campaigns were partly motivated by the continued efforts to stop plague from becoming an endemic problem in the American West. Since 1906, antiplague efforts in California had begun to devise techniques of squirrel extermination that utilized the capacities of fumigation devices and substances in the underground burrow systems of the ground squirrel. Best applied in moist conditions, when the soil's capacity to hold large amounts of gas were highest, the campaign officers adapted various devices from maritime sanitation.[74] One device was built in particular to enhance squirrel destruction by means of either poisoning the animals with gas or by injecting an inflammable gas, which would then be ignited to kill all squirrels within the extended burrow systems. The device, called the "squirrel destructor," was developed within the US Marine Hospital Service and used throughout the extensive campaigns in California from 1906; it was utilized in the 1924 plague outbreak in Los Angeles and remained in use until at least the 1940s.[75]

The importance of HCN became all the more pressing after the 1926 International Sanitary Conference in Paris, where it was decided that all ships should have a certificate of rat destruction or that they have been inspected and found to be rat-free within the last six months. These were known as the Deratization Certificate and the Deratization Exemption Certificate. The relevant article of the convention (Article 28) decreed that if neither certificate could be procured, then the port authorities could themselves carry out or direct the carrying out of deratization. No specific technology was singled out, but it was stated that "It shall decide in each case the technique which should be employed to secure the practical extermination of rats on board, but details of the deratizing process applied and of the number of rats destroyed shall be entered on the certificates."[76]

The increasing preference of cyanization over suplhurization is furthermore evident in the first and second International Rat Conference. During the first conference (May 16–22, 1928), the two deratization methods were discussed as on par with regard to their efficacy, with authors like Colombani (director of public health in the French protectorate of Morocco) and Herminier (director of the School of Health Service for colonial troops in Marseilles) noting the relative advantages of the two methods, and generally leaning toward the well-tested method of sulphurization, as this demonstrated a guaranteed safety for operating staff and had a germicidal property lacking in HCN.[77] At the same time, HCN was already recognized by some delegates with respect to the number of rats it killed, its rapidity of action, and its nondamaging properties of metals. The Italian delegate, Lutratio, noted that, as a result, use of HCN in Italian ports had increased in the past five years from 26 to 46 percent.[78] In the course of the conference, delegates were invited to witness, on May 20, 1928, the demonstration of different maritime deratization techniques and technologies in the harbor of the Havre. Though the demonstration included the Clayton and the Marot, its main focus was a new HCN-based apparatus, which had only recently been approved for use in France: the Sanos generator (*Sanos générateur*).[79]

The August 8, 1929, Decree on the Deratization of Vessels of the French Republic fully authorized the use of HCN for maritime deratization, on the provision that: a) this was accompanied by a gas detector; b) proscribed procedures were strictly followed; c) a declaration by the ship's captain was provided, taking responsibility that during fumigation with HCN nobody is on board with the exception of deratization staff and the relevant health officials.[80] By the time of the Second International Rat Conference, held in Paris on October 7–12, 1931, little doubt remained over the superior efficacy of HCN. Though sulphurization still featured in the discussions, HCN enjoyed the unambiguous support of a key player: large navigation companies, like the Compagnie Générale Transatlantique, whose chief doctor, Chamaillard, provided a glowing review of cyanization.[81]

The employment of HCN led to a proliferation of cyanide-based fumigation machines and it enjoyed great popularity and success across the globe in the decades to follow.[82] A manual on the protection of Ceylon (Sri Lanka) from plague from 1931 points to the reasons for this.[83] Thoroughly informed about US studies and publications regarding the fumigant, in his review of the use of HCN fumigation in the Sri Lanka highlands, the City Microbiologist of the colonial capital of the island, Colombo, L. F. Hirst, reserved particular praise for one apparatus invented in British India, the Liston Cyanide Fumigator:

> Solutions of sodium cyanide and acid are run into the lead-lined generating box and the gas thus generated is diluted with air drawn, by means of a petrol driven fan, through the generating box, thence by means of flexible gas trunks to the compartment to be fumigated and back to the machine. The mixture of air and

gas in the compartment is continually circulated. Samples of air can be readily drawn off from the outlet pipe and the concentration of HCN in the circulating gas speedily determined by Liebig's method or one of its modifications.[84]

Having been tried experimentally in lighters in Colombo's harbor, as well as in the laboratory, the advantage of this method comprised in it being performed in the open air, thus minimizing the risk of accident to the operating personnel. However, the small capacity of the machine required multiple apparatuses to operate in any given harbor. Liston's apparatus had been promoted by its inventor as an alternative to the "dumping fixture" method that had enjoyed particular success in the hands of Creel in the United States. This was because, according to Liston, "dumping fixtures should be avoided, not only because they must be placed within the space to be treated, but also because the gas, when generated in this apparatus, is evolved with almost explosive rapidity so that high, and therefore dangerous, concentrations are developed."[85] "The necessity for handling the corrosive and poisonous waste in a closed space," claimed Liston, "is an additional drawback to the use of the dumping fixture."[86]

The Ceylon report also mentioned another method of employing hydrocyanic acid for deratization. This involved a "German product," which consisted of "an infusorial earthy substance called diatomite absorbing about an equal weight of liquid HCN plus various amounts of chloropicrin (10 per cent., 6 per cent., or 4 per cent)"; this was meant to enable "the useful post-fumigation warning lachrymatory effect."[87] Sold in cans of various sizes, upon opening "the material can readily be poured out of the can into the compartment to be fumigated." The benefits of this new substance were briefly praised in the Ceylon report, though lack of detail suggests Hirst had no first-hand experience of its employment: "Owing to the great surface area of diatomite, evolution of the absorbed cyanide is very rapid and so complete that at the end of the customary two hour fumigation period the residue is therefore, be safely left *in situ* though it is advisable to sweep it up in order to obviate persistence of the chloropicrin odor and its tear effect."[88] The substance was no other than Zyklon B.

As Paul Weindling has demonstrated, historical scholarship tends to oversee the detailed and extensive history of the substance before it was applied in the gas chambers of the Holocaust. Indeed, developed as an improved derivative of hydrocyanic acid, the personnel at the Kaiser Wilhelm Institutes in Berlin had developed this commercially viable form "as a spin-off from defensive and offensive poison gas research under the chemist Fritz Haber."[89] The substance was essentially introduced into the market in 1915 as an easy-to-use variant of the acid, for application as a disinfectant in louse-infested compartments. Since 1917, the German and Austro-Hungarian military ran experimental research programs for the improvement of delousing procedures and pest control. Here, as well as in the United States, the gas was seen to be cheap, as well as harmless to fabrics, leather, and metal parts on uniforms, but it was also experimentally shown to be able to penetrate fabrics in all kinds of circumstances. Since April 1917, Weindling reports, the military had used hydrocyanic acid to disinfest over twenty-one million cubic meters of buildings by 1921.[90] But a number of incidents, reports of injuries and deaths prompted the further development of the substance to make it less dangerous to humans. Further complicated through the ban of "the use of asphyxiating, poisonous and other gases and all analogous liquids" in Germany (article 171 of the Treaty of Versailles), it took the invention of a further derivate and the addition of chlorine and bromide by the chemist Bruno Tesch, to finally arrive at Zyklon B in 1923: "Zyklon was thus endorsed as protecting public safety, overcoming the risks of hydrocyanic acid which continued to result in fatal accidents."[91]

Tesch, in collaboration with the Hamburg-based harbor physician Nocht, developed a code of conduct for disinfection works, and laid the groundwork for the special category of disinfectors—personnel trained to safely conduct the gasing with Zyklon B. Gas chambers were installed and refined throughout the 1920s for the purpose of commercial, occupational, and experimental disinfestation. Zyklon B's employment for deratization was not limited to Germany. The substance was also extensively tested in the United States.[92] An influential experiment on the deratization properties of cyanogenic fumigants led by Surgeon C. V. Akin and Acting Assistant Surgeon G. C. Sherrard in the New York Quarantine Station, unambiguously signaled that, by the late 1920s, maritime fumigation's goal was no longer disinfection in the sense of germicide, but disinfestation: "To kill rats is the prime object of ship fumigation."[93] It also pointed out the end of the era of sulphuric fumigation and the realization of its hygienic utopia by cyanide-based products instead:

> For stations not yet ready to relinquish sulphur as a fumigant, a useful combination will be found in sulphur for the holds and Zyklon B in all upper deck

compartments where the destructive effects of sulphur are objectionable. Such stations should be encouraged in the use of cyanide, however, as a sulphur fumigation is time consuming and, except in the instance of unusually well-prepared vessels, does not compare with cyanide.[94]

Similarly, at the Second International Rat Conference, Chamaillard seemed confident about the efficacy of the substance, which had already been used to some acclaim as far back as 1928 in the port of Rotterdam.[95] However, Zyklon B was not as conclusive as perhaps its manufacturers would have hoped. For example, in a report published in the United States in 1931, Surgeon C. L. Williams admitted that "it was reluctantly, and only after considerable experimentation, that the writer turned from Zyklon back to liquid HCN for the fumigation of vessels which are either loaded with cargo or have protected rat harborages which are heavily infested."[96] This was because unless the discoids were carefully scattered on the 'tween decks, Zyklon B failed to acquire an adequate concentration in the far corners of the holds, which were believed to be preferable areas for "rat harborage."[97] Such distribution required staff to place the discoids manually, which Williams claimed they were generally unwilling to do. By contrast, liquid HCN could be introduced by an air-jet spray apparatus that could allow precise manipulation, and, by means of the pressure provided by the air jet (at 200 pounds), reach even the remotest corners of the holds (figure 7.2). Comparing the use of this apparatus on loaded ships to Zyklon B, Williams noted that it was difficult to acquire accurate doses of the gas by means of the latter.

Moreover, when it came to reaching rats scurrying away, the spray method provided an unprecedented advantage: "Where there exist rat harborages, into which it is unlikely that the gas will itself penetrate in lethal concentration, the nozzle of the gun may be passed through small openings and the spray projected directly into them, securing a penetration far deeper than could be expected by any other means now in use."[98] Accordingly, Zyklon B should only be used for routine fumigations, and in smaller stations, which could not use compressed-air apparatuses or lacked personnel trained in its use.

In 1931, Surgeon J. R. Ridlon of the US Public Health Service in San Francisco confidently declared that "the use of suitable cyanogen products has practically replaced the use of sulphur at all of the quarantine stations of the larger ports."[99] The use of cyanogen products thus marked

FIGURE 6.—Projecting HCN spray directly into rat-infested insulation of a cold-storage room. The apparatus shown is the first one used, in which the handle and valve were assembled separately. A strip of casing has been removed to permit direct application of fumigant to insulated space

Figure 7.2

Projecting HCN spray directly into rat-infested insulation of a cold-storage room.
Source: C. L. Williams, "The air jet hydrocyanic acid sprayer," *Public Health Reports (1896–1970)* 46, no. 30 (July 24, 1931): plate II, figure 6.

the beginning of a gradual fumigation by means of large generators by individually held air-jet sprays that could directly "inject" fumigants into rat harborages. This followed a new doctrine in maritime sanitation, which, in Williams's words, relied on the fact that "it has been proved beyond the possibility of doubt that the mere release of a fumigant in an enclosed space does not insure penetration of the gas in lethal concentration in all retired locations and dead air spaces."[100]

Conclusion

The history of sulphuric utopias does not present a tale of a radical hygienic vision transformed into reality by means of a new technology. Nor is this merely an account of a failed technology, which never achieved the full abolishment of involuntary ship and cargo detention. Instead, we have undertaken the history of the development, production, dissemination, and experimental examination of a mechanical and chemical process, so as to think about sulphuric utopias as a focal point in which many histories of modernization and hygienic innovation coalesce.

First of all, the mobilization of sulphur-based fumigation relied on a long history of sulphurous alchemy both for therapeutic and hygienic purposes. Confronted with the requirements of quarantine and catalyzed by the conceptual foundations of disinfection, the vapor was reinvented. Sulphur was known for centuries to be therapeutic in supporting bodily balance and in providing a fortified atmosphere against pathogenic ailments. In the late nineteenth century, it was slowly transformed into a chemical gas with dedicated effects on organic forms of life. This is then a story of a chemical compound, supported by the budding science of bacteriology, in which the detrimental effect of SO_2 to microbes was observed and experimentally confirmed. Moreover, through the material requirements of maritime sanitation, sulphur-based fumigation became a practice attached to more than just the practicability of enabling hygienic cleanliness in hospitals or laboratories. At the end of the nineteenth century, a traditional compound was renewed as a chemical substance with antibacteriological capacities and applied through machines to fortify and encourage the belief in a continuous flow of trade without the risk of disease transmission.

Second, the technological advancement of disinfection intersects with the history of maritime trade and the late nineteenth-century push for economic globalization. We have shown that the concern about global shipping routes as pathways of epidemic disease provided a privileged site for the experimental implementation of chemically intrusive routines and practices. The development of maritime fumigation followed on from a drastic price reduction in global sea transport. The transport of basic goods, such as grains, tobacco, and cotton, from new territories in the colonies began to appear as an economic alternative to their production in Europe and the United States. Mostly attributed to the revolutionary transformations in shipping, the surge in global trade relied fundamentally on seamless, uninterrupted, and safe circulation of vessels and their goods. From Holt's invention in New Orleans to Calmette's considerations in experiments conducted in France and its territories, fumigation was seen as essential to guaranteeing the free flow of goods. The apparatus and its sulphuric potential were supposed to remove the persisting risk of epidemics from this newly emerging global economic order under European and American direction.

But when brought into association with the global archipelago of quarantine stations and islands, the ship, with its moving and suspended space, also set a scene for a different kind of hygienic utopia—one in which the capitalist dream of uninterrupted commerce and the colonial imaginary of uninhibited expansion became tied to visions of control over infectious diseases. This was the same vision that the laboratories of French and German bacteriology had enabled at the time. The utopia that fumigating machines embodied and catalyzed was one of the abolition of maritime disease transmission from global colonial and capitalist maps. But perhaps more importantly, its purpose was the protection and indeed defense of Europe and the United States against the importation of diseases that were seen as essentially foreign and belonging to a premodern or "tropical" chronotope.

The fumigation apparatus was thus positioned at the heart of the well-studied late nineteenth-century political nexus of infectious diseases, trade, and migration. Clayton and its variants collapsed three different frontiers: First, the device served as a barrier to bacterial and animal agents of epidemic diseases, separating spaces of infection from hygienic spaces of imagined health and salubrity. Second, the machine's installation in ports integrated this bacteriological barrier with the borderline between land and sea, and between the ship as carrier and the land behind the port as a vulnerable space. Thirdly, this biopolitical border regime was then folded into the colonial and imperial order as it doubled as a boundary regulation of the hygienic West, separating colonial space (the tropics) as one ontologically linked to disease and infection. We have shown that the securitization of this multiple borderline remained always fragile and inconsistent, for the mechanical contraption of the Clayton apparatus failed to secure these border regimes on microbiological, geographical, and political lines, as much as it failed to bring about a sulphuric utopia.

But while the Clayton machine might have failed in its utopian project, it succeeded in the production of an extraordinary global sociotechnical ensemble among epidemiologists, bacteriologists, engineers, and sanitary officers. The Clayton enabled a technoscientific vision of the free flow of goods, and promised a future of maximized control over foreign pathogens. As such, the epidemiological-bacteriological-sanitary ensemble did not only invest in the development, testing, and distribution of a machine, but it also contributed substantially to the consolidation of the imagination of infectious diseases and to the installation of invisible adversaries to public health: first the pathogen and later the infectious insect or rodent. Installed in ports around the world, tweaked, improved, and repurposed for different pathogens and vectors, the system of maritime sanitation first advanced and materialized by the Clayton machine circled for about four decades around a constantly shifting target of infectious matter. Like any experimental system, it had to adapt, reformulate, and reshape to bring its object into focus. This system began with the fumigation of undefined infectious matter causing yellow fever, moved quickly to the destruction of pathogens and bacteria, before it shifted to the obsessive consideration and destruction of plague-carrying rats.

Sulphuric Utopias has focused on the experimental system that was erected around the Clayton apparatus so as to emphasize the conditions under which the machine was considered to enable a disease-free future. This setup did not only change in terms of the target of fumigation; throughout its development, the product of the apparatus—the Clayton gas—emerged in its own unique and changing composition. The gas became largely defined through its capacities: first of all, its disinfective qualities, which involved not only the ability to interrupt the growth of microbes in the moment of application but to also render microbes innate. Not only did these microbes not show any further growth in agar, they were also incapable of causing infections in inoculated animals. Second, the sulphur-based gas was shown to have lethal effects on insects and rodents. A practice of disinfestation could thus be carried out in parallel and in combination with disinfection, rendering the Clayton machine and its gaseous product a versatile apparatus able to counter some of the key pathways in which infectious diseases were believed to be spreading. In other words, the apparatus brought into focus an understanding of infection in strong causal relationship to spaces, objects, pathogens, and vectors, thus promising a regime of complete and combined disinfection, disinsectization, and deratization.

Sulphuric Utopias has proposed an understanding of the broad applicability of the Clayton machine against all imaginable and intelligible infection routes at the turn of the nineteenth century as its key utopian quality. Not only did the Clayton promise an active compound that (re)turned ships and their goods into an inconspicuous state, but it also articulated and demonstrated a human capacity of acting against a whole range of organisms, which had been identified as antagonistic and indeed detrimental to human life.

There is, however, a paradox at the end of the sulphuric utopia outlined in this book: parallel to the radicalization of fumigation into a weapon of mass destruction and a technology of genocide (as ultimately exemplified in the Shoah), we also see a weakened expectation toward the capacities of what fumigation was supposed to achieve in maritime sanitation. As the Clayton machine and its competitors (both sulphur and carbon based) became the focus of experimental systems across the globe, they also became entangled in international debates about the very goal of fumigation (this was especially pronounced in international sanitary conferences and conventions). Maritime fumigation technologies transformed cargo boats into containers where tests were conducted in situ with the aim of reaching a standardized chemical and engineering process. The common goals were disinfection and disinfestation performed: a) across the maximum amount of space in the boat itself (rather than only near the floor of fumigated compartments, as was the case with heavy gases); b) at a maximum degree of penetration in the boat's structures, as well as the cargo itself; c) at a minimum risk for operators; d) at a minimum damage to both goods and the boat's structures; and e) at a minimum time of operation. As the hope of reaching such a "sulphuric utopia" became time and again entangled in the webs of political, economic, scientific, and imperial competition and antagonism, the very conditions of achieving the goal of standardized "disinfection plus disinfestation" came to transform the standard in question itself: the goal of disinfection was eventually given up, to focus instead on disinfestation, the eradication of pests.

With the introduction of HCN, germs lost their status of being the principle target of fumigation. The rapid success of HCN and the pervasive global replacement of sulphur-based fumigation systems after WWI was accompanied by the acknowledgement of HCN's relative inefficiency against bacteria. Cyanide-based fumigation gases seemed to leave most bacteria unharmed, while exhibiting an unparalleled efficiency in the destruction of living organisms. As the birth of the Clayton machine was conceived around the paradigm of disinfection (the destruction of bacteria), its end was marked by a systematic replacement of the goal of radical disinfection with that of the eradication of rats and the disinfestation of ships. This shift in the criteria of the efficiency and the very aim of fumigation aligns with a rising pragmatism in the fight against infectious diseases. This pragmatism was entangled with a simultaneously growing insight into the complexity of epidemics after WWI, when monocausal models began to be challenged by more ecological concepts of infectious diseases.

With the demise of sulphur came the end of sulphuric utopia as a vision of total disinfection, disinsectization, and deratization. As shown in this book, sulphuric acid had been lauded from its early systematic application in the 1880s as a chemical compound capable of eradicating germs and insects, as well as rodents. Persistent ambiguities about the role of insects in the transmission of yellow fever, and the ongoing controversy about the role of the rat in the case of plague reshaped the experimental system of the Clayton machine and its rivals up until the second decade of the twentieth century. While bacteriology was often charged with the introduction of a "biological reductionism" in epidemiological reasoning, the Clayton owed much of its success to the declared "war on germs." Yet, at the same time, we have shown that it also sustained a conceptual dedication to infection as a nonspecific process, implicating the material conditions that foster bacteria as well as their hosts and vectors.

Sulphuric Utopias has demonstrated that after the conceptual emergence of the Clayton in New Orleans, the apparatus was increasingly integrated in the global emergence of the rat as the central adversary to public health and maritime hygiene. While the paradigm of disinfection became quickly stabilized, and SO_2 was shown to provide conclusive results in the destruction of bacteria, its capacity in relation to rats was not similarly trusted. In competition with carbon-based gases, and embedded in international conflicts with the Ottoman Empire, the rat appeared as a new frontier against which the Clayton had to prove itself. What followed was a series of ship-based experiments and international exchanges that eventually established the Clayton as a standardized and globally recognized procedure, but without a universal consensus about its deratization capacities ever being reached. Particularly illuminating is the Argentinean episode, where an electrified version of SO_2, produced by the Marot apparatus, was believed to have such overarching success against rats and other rodents, that the technology and ideology of the machine was transported from the seafront and was integrated into the fabric of the city itself, by means of fulfilling the unique Argentinean version of sulphuric utopia as "general prophylaxis."

The displacement of the holy trinity of disinfection, disinsectization, and deratization by disinfestation as the goal of maritime sanitation was, however, what would eventually lead to the demise of the Clayton and other sulphur-based technologies. Port authorities' increasing confidence with hitherto-considered too dangerous HCN-based fumigation, as well as HCN's unambiguous capacity as regards rat and insect extermination, would set aside sulphur as a second-choice chemical. The ability to spray HCN directly into crevices inside ships' holds also held another advantage over the Clayton, for it could reach spaces allegedly out of bounds for sulphuric gases. This aspect of HCN fumigation would lead the way, after the end of World War II, for the transformation of vector control via the employment of hand-held sprays, using a new highly effective substance: DDT.

Working in tandem with other methods of controlling vectors of infectious diseases, such as rat-proofing, these technological transformations in maritime sanitation formed part of a new era of hygienic modernity whose aim was less about total disinfection and more about creating and maintaining barriers and distances (physical as well as chemical) between humans and infectious disease vectors. At the same time, beyond the realm of infectious diseases, they led to applications already prefigured by the military use of gases in WWI and chloropicrin in the Rif War in the 1920s. These reached their nadir in the way in which the technoscientific frame of fumigation played into the ideologies of Nazi Germany, as the methods, metaphors, and experimental setup were appropriated by the SS to utilize Zyklon B in the gas chambers of Auschwitz.

A defining mark of modernity has been the entanglement of machines and utopian visions in phantasmagorias of liberation and progress. The maritime fumigation technologies discussed in this book show that this connection goes beyond the locus of application preferred by most critical theorists, anthropologists, and historians: the city. As a floating but at the same time moorable space, ships formed a historically vital if analytically neglected ground for a global machinic travail toward a uniquely capitalist and imperialist utopia: the liberation of trade not only from infectious diseases, but also from the time restraints imposed on it by the latter in the form of quarantine.

Notes

Introduction

1. For the quarantine procedures and a detailed overview of the system of maritime sanitation, as devised by Joseph Holt, see Joseph Holt, "Quarantine operations," in *Annual Report of the Louisiana Board of Health, 1885 and 1886*, ed. Louisiana Board of Health (New Orleans: Louisiana Board of Health 1887), 23–38. See also Joseph Holt, "The sanitary protection of New Orleans, municipal and maritime," *The Sanitarian*, no. 194 (1886): 37–49.

2. Donna A. Barnes, *The Louisiana Populist Movement, 1881–1900* (Baton Rouge: Louisiana State University Press, 2011), 67.

3. Norman Howard-Jones, *The Scientific Background of the International Sanitary Conferences, 1851–1938* (Geneva: World Health Organization, 1975); Peter Baldwin, *Contagion and the State in Europe, 1830–1930* (Cambridge: Cambridge University Press, 2005); Mark Harrison, *Contagion: How Commerce Has Spread Disease* (New Haven, CT: Yale University Press, 2013); Alison Bashford, *Imperial Hygiene: A Critical History of Colonialism, Nationalism and Public Health* (Basingstoke, UK: Palgrave Macmillan, 2004); Marcos Cueto, *The Value of Health: A History of the Pan American Health Organization* (Washington, DC: Pan American Health Organization, 2007).

4. Alison Bashford. ed., *Quarantine: Local and Global Histories* (London: Palgrave Macmillan, 2016).

5. An overview of these systems can be gleamed from Anne Hardy, *The Epidemic Streets: Infectious Disease and the Rise of Preventive Medicine, 1856–1900* (Wotton-under-Edge, UK: Clarendon Press, 1993); Michael Worboys, *Spreading Germs: Disease Theories and Medical Practice in Britain, 1865–1900* (Cambridge: Cambridge University Press, 2000); Christoph Gradmann, "A spirit of scientific rigour: Koch's postulates in twentieth-century medicine," *Microbes and Infection* 16, no. 11 (November 2014): 885–892; J. Andrew Mendelsohn, "From eradication to equilibrium. How epidemics became complex after World War I," in *Greater than the Parts: Holism in Biomedicine,*

1920–1950, ed. Christopher Lawrence and George Weisz, (Oxford: Oxford University Press, 1998), 303–334.

6. Edmund Russell, *War and Nature: Fighting Humans and Insects with Chemicals from World War I to Silent Spring* (Cambridge: Cambridge University Press, 2001).

7. Linda Nash, *Inescapable Ecologies: A History of Environment, Disease, and Knowledge* (Berkeley, CA: The University of California Press, 2006).

8. William Coleman, *Yellow Fever in the North: The Methods of Early Epidemiology* (Madison: University of Wisconsin Press, 1987); François Delaporte, *The History of Yellow Fever: An Essay on the Birth of Tropical Medicine*, trans. Arthur Goldhammer (Cambridge, MA: MIT Press, 1991); Margaret Humphreys, *Yellow Fever and the South* (Baltimore, MD: Johns Hopkins University Press, 1999).

9. Carroll and Reed were part of the Cuba Yellow Fever Commission, which in 1900 proved experimentally that the mosquito *Aedes Aegypti* was indeed the primary vector of yellow fever, a theory that had been first conceptualized by Carlos Finlay in 1881. The results of the experiments were showcased in a large-scale and widely successful fumigation campaign across Cuba. José Amador, *Medicine and Nation Building in the Americas, 1890–1940* (Nashville, TN: Vanderbilt University Press, 2015), 30ff. See also Jaime Larry Benchimol, *Dos micróbios aos mosquitos: febre amarela e a revolução pasteuriana no Brasil* (Rio de Janeiro: SciELO—Editora FIOCRUZ, 1999).

10. Myron Echenberg, *Plague Ports: The Global Urban Impact of Bubonic Plague, 1894–1901* (New York: New York University Press, 2007).

11. Hans-Jörg Rheinberger, "Experimental systems. Historiality, narration, and deconstruction," in *The Science Studies Reader*, ed. Mario Biagioli (London and New York: Routledge, 1999), 417–429.

12. Worboys, *Spreading Germs*; Nash, *Inescapable Ecologies;* Mark Honigsbaum, "'Tipping the balance': Karl Friedrich Meyer, "Latent infections, and the birth of modern ideas of disease ecology," *Journal of the History of Biology* 49, no. 2 (2016): 261–309.

13. Indeed, as Graham Mooney points out, the invention of classical and even Homeric progenitors of fumigation was part and parcel of the latter's Victorian consolidation. Graham Mooney, *Intrusive Interventions: Public Health, Domestic Space, and Infectious Disease Surveillance in England, 1840–1914* (Rochester, NY: University of Rochester Press, 2015).

14. Hans-Peter Kroner, "From eugenics to genetic screening. Historical problems of human genetic applications," in *The Ethics of Genetic Screening*, ed. Ruth F. Chadwick, Darren Shickle, H. A. Ten Have, and Urban Wiesing (Berlin: Springer Science, 1999), 131–140; Lars Bluma, "The hygienic movement and German mining, 1890–1914," in *History of the Workplace: Environment and Health at Stake*, ed. Lars Bluma and Judith Rainhorn (London and New York: Routledge, 2015), 7–26.

15. Baldwin, *Contagion and the State*, 227–228.

16. For discussion of utilitarian and Marxist receptions of utopianism, see Ruth Levitas, *The Concept of Utopia* (Oxford: Peter Land, 2010).

17. See: Charles E. Rosenberg, "The therapeutic revolution: Medicine, meaning and social change in nineteenth-century America," *Perspectives in Biology and Medicine* 20, no.4 (1977): 485–506; Allan M. Brandt, *No Magic Bullet: A Social History of Venereal Disease in the United States since 1880* (New York: Oxford University Press, 1987).

18. Mark Olseen, "Totalitarianism and the 'repressed' utopia of the present: Moving beyond Hayek, Popper and Foucault," *Policy Futures in Education* 1, no. 3 (2003): 526–552.

19. Ernst Bloch, *The Principle of Hope*, trans. Neville Plaice, Stephen Plaice, and Paul Knight (Cambridge, MA: MIT Press, 1995).

20. Christos Lynteris and Branwyn Poleykett, "The anthropology of epidemic control: technologies and materialities," *Medical Anthropology* 37, no. 6 (2018): 433–441.

21. Ruth Levitas, "Educated hope: Ernst Bloch on abstract and concrete utopia," *Utopian Studies* 1, no. 2 (1990), 14.

22. Ibid., 15.

23. Bloch in Ibid., 14.

24. E. P. Thompson, *William Morris: Romantic or Revolutionary?* (London: The Merlin Press, 2011).

25. Ibid., 18, 17.

26. On "dreamscapes of modernity," see S. Jasanoff and S.-H. Kim (eds.), *Dreamscapes of Modernity: Sociotechnical Imaginaries and the Fabrication of Power* (Chicago: University of Chicago Press, 2015). On the relation between Bloch's notion of hope and the production of knowledge, see Hirokazu Miyazaki, *The Method of Hope: Anthropology, Philosophy, and Fijian Knowledge* (Stanford, CA: Stanford University Press, 2004).

27. Cat Moir, "Casting a picture. Utopia, Heimat and the materialist concept of history," *Anthropology and Materialism* 3 (2016), https://am.revues.org/573; see also *Utopian Studies* 25, no. 1, Special Issue: Utopia and Architecture (2014).

28. Reinhart Koselleck, *The Practice of Conceptual History: Timing History, Spacing Concepts* (Stanford, CA: Stanford University Press, 2002).

29. Sebastian Conrad, *Globalisation and the Nation in Imperial Germany*, (Cambridge: Cambridge University Press, 2010), 8ff.

30. Kevin H. O'Rourke and Jeffrey G. Williamson, *Globalization and History: The Evolution of a Nineteenth-Century Atlantic Economy* (Cambridge, MA: MIT Press, 1999), chapter 3. Economic historians tend to ignore the impact of epidemics, hygiene, and technologies of sanitation in their account of the history of global trade.

31. Douglas Burgess, *Engines of Empire: Steamships and the Victorian Imagination* (Stanford, CA: Stanford University Press, 2017).

32. Tom Koch, *Disease Maps: Epidemics on the Ground* (Chicago: University of Chicago Press, 2011). Mark Monmonier, "Maps as graphic propaganda for public health," in *Imagining Illness: Public Health and Visual Culture*, ed. David Harley Serlin (Minneapolis: University of Minnesota Press, 2010), 108–125.

33. Michel Foucault in Burgess, *Engines of Empire*, 12. See also: Michel Foucault, "Of Other Spaces," *Diacritics* 16, no. 1 (1986): 22–27. https://doi.org/10.2307/464648.

34. David Boyd Haycock and Sally Archer (eds.), *Health and Medicine at Sea, 1700–1900* (Woodbridge, UK: Boydell Press, 2009).

35. Burgess, *Engines of Empire*, 17.

36. Pierre Bouguer, *Traité du navire, de sa construction, et de ses mouvemens* (Paris: Jombert, 1746).

37. Tamson Pietsch, "A British sea: making sense of global space in the late nineteenth century," *Journal of Global History* 5, no. 3 (November 2010): 423.

38. Alison Bashford, "Maritime quarantine: Linking Old World and New World histories," in *Quarantine: Local and Global Histories*, ed. Alison Bashford (London: Palgrave Macmillan, 2016), 11.

39. Krista Maglen, *The English System: Quarantine, Immigration and the Making of a Port Sanitary Zone* (Manchester, UK: Manchester University Press, 2014), 7.

40. Michel-Rolph Trouillot, *Silencing the Past: Power and the Production of History* (Boston, MA: Beacon Press, 2015).

41. Elizabeth Kolbert, "Louisiana's disappearing coast," *New Yorker*, March 25, 2019, https://www.newyorker.com/magazine/2019/04/01/louisianas-disappearing -coast; John McPhee, "Atchafalaya," *New Yorker*, February 15, 1987, https://www .newyorker.com/magazine/1987/02/23/atchafalaya.

42. Erwin Ackerknecht, "Anticontagionism between 1821 and 1867," *Bulletin of the History of Medicine* 22 (1948): 562–593.

43. Ibid., 570; A similar constellation is described in detail for the case of New York in Howard Markel, *Quarantine!: East European Jewish Immigrants and the New York City Epidemics of 1892* (Baltimore, MD: Johns Hopkins University Press, 1999).

44. David S. Barnes, "Cargo, 'infection,' and the logic of quarantine in the nineteenth century," *Bulletin of the History of Medicine* 88, no. 1 (2014): 75–101.

45. Maglen, *The English System*, 7. On the importance of stasis as a medico-juridical notion, bridging human pathology and political disorder, see Kostas Kalimtzis, *Aristotle on Political Enmity and Disease: An Inquiry into Stasis* (New York: SUNY Press, 2000).

46. Miguel Abensour, "Persistent utopia," *Constellations* 15, no. 3 (2008): 407.

47. There are no thorough accounts of the historical conditions under which sulphuric and sulphurous acids became an epistemic object in the modern sciences. For oxygen as a comparable case, see Hasok Chang, "The persistence of epistemic objects through scientific change," *Erkenntnis* 75, no. 3 (November 2011): 413–429.

48. Simon Schaffer, David Serlin, and Jennifer Tucker, "Editors' introduction, Special Issue: The history of technoscience," *Radical History Review* 127 (January 2017): 3.

49. Simon Schaffer, "Measuring virtue: Eudiometry, enlightenment, and pneumatic medicine," in *The Medical Enlightenment of the Eighteenth Century*, eds. Andrew Cunningham and Roger French (Cambridge: Cambridge University Press, 1990), 281–318; Steven Shapin and Simon Schaffer, *Leviathan and the Air-Pump: Hobbes, Boyle, and the Experimental Life* (Princeton, NJ: Princeton University Press, 2011).

50. See, for example, Randall Packhard, *A History of Global Health: Interventions into the Lives of Other Peoples* (Baltimore, MD: Johns Hopkins University Press, 2016); Wenzel Paul Geissler (ed.), *Para-states and Medical Science: Making African Global Health* (Durham, NC: Duke University Press, 2015). David Reubi, "A genealogy of epidemiological reason: Saving lives, social surveys and global population," *BioSocieties* 13, no. 1 (March 2018): 81–102.

51. Marianne de Laet and Annemarie Mol, "The Zimbabwe bush pump: Mechanics of a fluid technology," *Social Studies of Science* 30, no. 2 (April 2000): 225–263.

52. Peter Redfield has more recently critically reapproached this "non-heroic vision of design" (2016: 160). Examining a postcolonial humanitarian device (the LifeStraw) Redfield puts stress on "science and technology in the form of infrastructure, the material frontline of norms that define 'modernity,' particularly in their absence" (Ibid). His analysis thus leads to an understanding of humanitarian devices in the age of neoliberalism as "substitute micro-infrastructures" deployed not simply in the absence of public (and public health) infrastructures but also in substitution of the latter's promissory and desiring regime: the commodified humanitarian pragmatics of "making a difference" for the emancipatory utopia of health as a public good. See Peter Redfield, "Fluid technologies: The bush pump, the LifeStraw® and microworlds of humanitarian design," *Social Studies of Science* 46, no. 2 (April 2016): 159–183.

53. De Laet and Mol, "The Zimbabwe Bush Pump."

54. Redfield, "Fluid technologies," 160.

55. De Laet and Mol, "The Zimbabwe Bush Pump," 252.

56. Gilles Deleuze and Felix Guattari, *Capitalism and Schizophrenia I: Anti-Oedipus*, trans. Robert Hurley (London: Penguin, 2009). See also Stephen J. Collier, Jamie Cross, Peter Redfield, and Alice Street (eds.), "Little development devices/humanitarian goods," *Limn* 9 (November 2017), https://limn.it/issues/little-development-devices -humanitarian-goods/.

Chapter 1

1. On the origin of the sulphur-mercury theory, see Allen G. Debus, *The English Paracelsians: The Chemical Challenge to Medical and Scientific Tradition in Early Modern France* (Cambridge: Cambridge University Press, 1991) and William R. Newman, "Alchemical and chymical principles. Four different traditions," in *The Idea of Principles in Early Modern Thought: Interdisciplinary Perspectives*, ed. Peter R. Anstey (London & New York: Routledge, 2017), 77–97.

2. James J. Garber, *Harmony in Healing: The Theoretical Basis of Ancient and Medieval Medicine* (London and New York: Routledge, 2008), 148.

3. Katherine Eggert, *Disknowledge: Literature, Alchemy, and the End of Humanism in Renaissance England* (Philadelphia: University of Pennsylvania Press, 2015), 65.

4. Claude Dariot in Allen G. Debus, *The Chemical Philosophy* (Mineola, NY: Dover Publications, 1977), 158.

5. For a concise history of mercurial treatments of syphilis, see Gérard Tilles and Daniel Wallach, "Le traitement de la syphilis par le mercure. Une histoire thérapeutique exemplaire," *Revue d'histoire de la pharmacie* 84, no. 312 (1996): 347–351.

6. Pierre Lalouette, *Nouvelle méthode de traiter les maladies vénériennes, par la fumigation: avec les procès-verbaux des Guérisons opérées par ce moyen* (Paris: Merigot, 1776); translation taken from Pierre Lalouette, *A New Method of Curing the Venereal Disease by Fumigation: Together with Critical Observations on the Different Methods of Cure and an Account of Some New and Useful Preparations of Mercury* (London: J. Wilkie, 1777), 12.

7. Ibid., 39

8. Ibid., 117–118.

9. On the camera obscura as a disciplinary apparatus, see Jonathan Crary, *Techniques of the Observer: On Vision and Modernity in the Nineteenth Century* (Cambridge, MA: MIT Press, 1990).

10. Halle, Dubois, Pinel, and Dupuytren in John Revere, *An Inquiry into the Origins and Effects of Sulphurous Fumigations in the Cure of Rheumatism, Gout, Diseases of the Skin, Palsy &c* (Baltimore, MD: Coale & Co, 1822), 5.

11. Ibid., 7.

12. Marcia Ramos-e-Silva, "Giovan Cosimo Bonomo (1663–1696): Discoverer of the etiology of scabies," *International Journal of Dermatology* 37, no.8 (August 1998): 625–630; M. A. Montesu and F. Cottoni, "G. C. Bonomo and D. Cestoni. Discoverers of the parasitic origin of scabies," *American Journal of Dermatopathology* 13, no. 4 (August 1991): 425–427.

13. Revere, *An Inquiry into the Origins and Effects of Sulphurous Fumigations.*

14. Halle et al., in ibid., 9.

15. Ibid., 10.

16. J. C. Galès, *Mémoires et rapports sur les fumigations sulfureuses appliquées au traitement des affections cutanées et de plusieurs autres maladies* (Paris: L'Impremerie Royale, 1816), 6.

17. Ibid., 7.

18. G. Emerson, "Cases illustrative of the efficacy of sulphurous fumigations, in certain cases—with preliminary remarks," *The Philadelphia Journal of the Medical and Physical Sciences* 3 (1821): 130.

19. Ibid., 132.

20. George Alfred Walker, *A Treatise on the Cure of Ulcers by Fumigation* (London: Longman, 1847), 75.

21. For their report, see Revere, *An Inquiry into the Origins and Effects of Sulphurous Fumigations.* Whether he should be acknowledged as the inventor of the method formed the subject of acrimonious public debate, with the doctor's critics claiming that since a sulphurization box had been mentioned by Glaubert in 1659, the modern box was simply an adaptation. The tribunal's opinion sided with those who considered that early modern invention to be forgotten and never applied in practice. For discussion, see Conseil général d'administration des hospices, *Descriptions des appareils à fumigations, établis, sur les dessins de M. D'Arcet, à l'hôpital Saint-Louis, en 1814 et successivement dans plusieurs Hôpitaux de Paris, pour le traitement des Maladies de la peau* (Paris: Impr. des Hospices civils, 1818).

22. Walker, *A Treatise on the Cure of Ulcers by Fumigation.*

23. Jean de Cerro, *Observations pratiques sur les fumigations sulfureuses (*Vienna: Charles Gerold 1819); David Luthy, *Notice sur les fumigations sulfureuses, appliquées au traitement des affections cutanées, et de plusieurs autres maladies, avec la description*

exacte d'un appareil pour les administrer (Fribourg, Switzerland: Aloyse Eggendorffer, 1818).

24. Mary Douglas, *Purity and Danger, an Analysis of Concepts of Pollution and Taboo* (London & New York, Routledge, 2002 [1966]). For studies following Douglas's reading of pollution, see Mark Bradley (ed.), *Rome, Pollution and Propriety. Dirt, Disease and Hygiene in the Eternal City from Antiquity to Modernity* (Cambridge: Cambridge University Press, 2012).

25. Jacques Jouanna, *Greek Medicine from Hippocrates to Galen: Selected Papers* (Leiden, Netherlans: Brill, 2012), 124.

26. For a discussion of *loimos*, and its difference, synergy, and confusion with *limos*, see Robin Mitchell-Boyask, *Plague and the Athenian Imagination: Drama, History, and the Cult of Asclepius* (Cambridge: Cambridge University Press, 2007); Rachel Bruzzone, "*Polemos, pathemata*, and plague: Thucydides' narrative and the tradition of upheaval," *Greek, Roman, and Byzantine Studies* 57 (2017): 882–909.

27. Jouanna, *Greek Medicine from Hippocrates to Galen*. See also: Fabian Meinel, *Pollution and Crisis in Greek Tragedy* (Cambridge: Cambridge University Press, 2015). For the invocation of this Homeric image as a progenitor of fumigation in Victorian texts see: Graham Mooney, *Intrusive Interventions: Public Health, Domestic Space, and Infectious Disease Surveillance in England, 1840–1914* (Rochester, NY: University of Rochester Press, 2015).

28. Galen, in the only passage of his work where he discusses miasmata, recounts Pliny's and Plutarch's legend that Hippocrates used fire using fragrant wood against the plague of Athens; no evidence of this practice exists in the Hippocratic corpus or in Thucydides. For discussion, see Jody Rubin Pinault, *Hippocratic Lives and Legends* (Leiden, Netherlands: Brill, 1992). On Hippocrates and gynecological fumigation, see Helen King, *Hippocrates' Woman: Reading the Female Body in Ancient Greece* (London and New York: Routledge, 1998).

29. Halle et al. in Revere, *An Inquiry into the Origins and Effects of Sulphurous Fumigations*, 6.

30. On plague and stench in medieval England, see Carole Rawcliffe, "'Great stenches, horrible sights and deadly abominations': Butchery and the battle against plague in late medieval English towns," in *Plague and the City*, eds. Lukas Engelmann, John Henderson, and Christos Lynteris (London and New York: Routledge, 2018), 18–38.

31. Joseph P. Byrne, *Encyclopedia of the Black Death* (Santa Barbara, CA: ABC-CLIO, 2012). The hybrid was already in place in Galen's plague of Athens anecdote; see note 28, this chapter.

32. Père Maurice de Toulon, *Le Capucin charitable enseignant le méthode pour remédier aux grandes misères que la peste a coutume de causer parmi les peoples* (Lyon, France: Bruyet: 1722). On the question of corpses as sources of infection across the centuries, see Christos Lynteris and Nicholas H. Evans (eds.), *Histories of Post-Mortem Contagion: Infectious Corpses and Contested Burials* (London: Palgrave Macmillan, 2017).

33. For an account and diagram of how the machine would have worked, see Sylvain Gagnière, *La désinfection des caveaux d'églises après les grandes épidémies de peste* (Avignon, France: Impr. Rulliere frères, 1943).

34. Ibid.

35. The Franque family was the most successful family of architects in eighteenth-century Avignon, famous for its hospital designs; Béatrice Gaillard, "Les Franque et les bâtiments hospitaliers d'Avignon au XVIIIe siècle: entre tradition et mutations," *In Situ: Revue des patrimonies* 31 (2017), http://insitu.revues.org/14242.

36. Gagnière, *La désinfection des caveaux d'églises après*, 10–11.

37. Steven Connor, *The Matter of Air. Science and Art of the Ethereal* (London: Reaktion Books, 2010).

38. James Lind, *Two Papers on Fevers and Infections Which Were Read before the Philosophical and Medical Society in Edinburgh* (London: D. Wilson, 1763).

39. Ibid., 45.

40. Ibid., 46.

41. Ibid.

42. Ibid., 46–47.

43. Ibid., 49.

44. James Lind, *An Essay on the Most Effective Means of Preserving the Health of Seamen in the Royal Navy* (London: D. Wilson, 1762), 120.

45. John Howard, *The State of the Prisons in England and Wales: With Preliminary Observations and an Account of Some Foreign Prisons* (Cambridge: Cambridge University Press, 2013 [1777]), 166.

46. Louis-Bernard Guyton de Morveau, *Traité des moyens de désinfecter l'air* (Paris: Bernard, 1801). Translations taken from Louis-Bernard Guyton de Morveau, *A Treatise on the Means of Purifying Infected Air*, transl. R. Hall (London: J. & E. Hodson, 1802).

47. Ibid., 27.

48. Ibid. On Lavoisier's impact on Guyton, see Alain Corbin, *Le miasma et la jonquille. L'odorat et l'imaginaire social aux XVIIIe et XIXe siècles* (Paris: Champs histoire,

2016 [1982]); Elana Serrano, "Spreading the revolution: Guyton's fumigating machine in Spain. Politics, technology, and material culture (1796–1808)" in *"Astonishing Transformations": How Chemistry Made and Managed the World, 1760–1840*, eds. Lissa Roberts and Simon Warrett (Amsterdam: Brill, 2017), 106–130.

49. Guyton de Morveau, *Traité des moyens de désinfecter l'air*, 28.

50. On Sydenham's concepts of epidemics, see Knut Faber, *Nosography. The Evolution of Clinical Medicine in Modern Times* (New York: AMS, 1930).

51. James Carmichael Smyth, *Description of the Jail Distemper as It Appeared amongst the Spanish Prisoners at Winchester, in the Year 1780* (London: J. Johnson, 1795).

52. Ibid., 7.

53. Ibid., 40.

54. Ibid., 40–41.

55. Ibid., 46.

56. Ibid., 54.

57. Ibid., 183–184.

58. "Report from the Committee on Dr. Smyth's Petition, respecting his Discovery of Nitrous Fumigation, Reported by Henry Bankes, Esquire, 10th June 1800," in *Reports from the Committees of the House of Commons, Reprinted by Order of the House*, Vol. XIV, Miscellaneous Reports 1793–1802, 192.

59. Ibid.

60. Ibid., 195.

61. Ibid., 193.

62. Ibid., 198.

63. On life and work of Louis Odier, see de Georges Morsier, "La vie et l'œuvre de Louis Odier, docteur et professeur en médecine (1748—1817)," *Gesnerus* 32 (1975): 248–270.

64. On the parliamentary process, see Catherine Kelly, "Parliamentary inquiries and the construction of medical argument in the early 19th century, 1793–1825," in *Lawyers' Medicine: The Legislature, the Courts and Medical Practice, 1760–2000*, eds. Imogen Goold and Catherine Kelly (London: Bloomsbury, 2000).

65. "Report from the Committee on Dr. Smyth's Petition," 189.

66. Ibid., 189.

67. Guyton de Morveau, *Traité des moyens de désinfecter l'air*.

68. Serrano, "Spreading the revolution."

69. Ibid. For a study of the impact of Guyton's fumigation method in the Netherlands, see Teunis Willem van Heiningen, "La contribution à la santé publique de Louis-Bernard Guyton de Morveau (1737–1816) et l'adoption de ses idées aux Pays-Bas," *Histoire des sciences médicales* XLVIII, no. 1 (2014): 97–106. For a manual comparing Guyton's method with other fumigation methods, and their efficacy, see Carlos de Gimbernat y Grassot, *Instruction sur les moyens propres à prévenir la contagion des fièvres épidémiques* (Strasbourg, France: F. G. Levrault, 1814).

70. As Serrano notes, this machine may look simple by comparison to previously described fumigators but it in fact incorporated extensive material and technological research as well as craftsmanship.

71. Serrano, "Spreading the revolution," 106.

72. Ibid., 116, in the original the quote is all in upper case.

73. Ibid., 117.

74. Ibid., 120.

75. Antonio García Belmar and Ramón Bertomeu-Sánchez, "L'Espagne fumigée. Consensus et silences autour des fumigations d'acides minéraux en Espagne, 1770–1804," *Annales historiques de la Révolution française* 383, no. 1 (2016): 177–202.

76. As Serrano (2017: 123) argues, fumigation allowed Miguel Cabanellas to design a new lazaretto model which may indeed "be understood as a chemically-based plant for recycling people and goods back to normal circulation." See: Miguel Cabanelles, *Defensa de las fumigaciones ácido-minerales* (Madrid: Repullés, 1814). This use of fumigation for perfecting quarantine is also evident in the uses of Guyton's method in the Netherlands; Heiningen, "La contribution à la santé publique de Louis-Bernard Guyton de Morveau."

77. Friedrich Schnurrer, "Die geographische Verteilung der Krankheiten, vorgelesen in der Versammlung der deutschen Aerzte und Naturforscher zu München den 22. Sept. 1827," *Das Ausland* 1 (1828): 357–359; Friedrich Schnurrer, *Die Cholera Morbus, ihre Verbreitung, ihre Zufälle, die versuchten Heilmethoden, ihre Eigenthümlichkeiten und die im Großen dagegen anzuwendenden Mittel* (Stuttgart, Tübingen: Gotta'sche Buchhandlung, 1831); R. Broemer, "The first global map of the distribution of human diseases: Friedrich Schnurrer's 'Charte Über Die Geographische Ausbreitung Der Krankheiten' (1827)," *Medical History* Supplement 20 (2000): 176–85.

78. Köslin (Regierungsbezirk), *Amts-Blatt der preußischen Regierung zu Köslin* (Köslin, Prussia: C. G. Hendess, 1831), 172, 327.

79. H. Scoutetten, *Relation historique et médicale de l'épidémie de choléra qui a régné à Berlin en 1831* (Paris: J.-B. Baillière, 1832).

80. Erwin H. Ackerknecht, "Anticontagionism between 1821 and 1867," *Bulletin of the History of Medicine* 22 (1948): 562–593; Howard Markel, *Quarantine!: East European Jewish Immigrants and the New York City Epidemics of 1892* (Baltimore, MD: Johns Hopkins University Press, 1999); Krista Maglen, *The English System: Quarantine, Immigration and the Making of a Port Sanitary Zone* (Manchester, UK: Manchester University Press, 2014).

81. Guenter B. Risse, *Driven by Fear: Epidemics and Isolation in San Francisco's House of Pestilence* (Champaign, IL: University of Illinois Press, 2015).

82. John Henderson, "Filth is the mother of corruption: plague and the built environment in early modern Florence," in *Plague and the City*, eds. Lukas Engelmann, John Henderson, and Christos Lynteris (London and New York: Routledge, 2018), 69–90.

83. Markel, *Quarantine!*. See also Alan M. Kraut, *Silent Travelers: Germs, Genes, and the "Immigrant Menace"* (New York: BasicBooks, 1995); Nayan Shah, *Contagious Divides: Epidemics and Race in San Francisco's Chinatown* (Berkeley, CA: University of California Press, 2001).

84. Isabell Lorey, *Figuren des Immunen. Elemente einer Politischen Theorie* (Zürich: Diaphanes, 2011); David Napier, *The Age of Immunology: Conceiving a Future in an Alienating World* (Chicago, IL: The University of Chicago Press, 2003).

85. Mark Harrison, *Contagion: How Commerce Has Spread Disease* (New Haven, CT: Yale University Press, 2013).

86. Ibid., 5

87. Alison Bashford, *Imperial Hygiene. A Critical History of Colonialism, Nationalism and Public Health* (Basingstoke, UK: Palgrave Macmillan, 2004).

88. Ackerknecht, "Anticontagionism between 1821 and 1867."

89. Harrison, *Contagion*.

90. Perrier, cited in Norman Howard-Jones, *The Scientific Background of the International Sanitary Conferences, 1851–1938* (Geneva: World Health Organization, 1975), 14.

91. Ibid., 35. For a discussion of the Ottoman quarantine system, see chapter 3.

92. Howard-Jones, *The Scientific Background of the International Sanitary Conferences, 1851–1938*, 35.

93. Ibid., 65. On early twentieth-century debates on this classification and its relation to fumigation, see chapter 4.

94. Myron Echenberg, *Plague Ports: The Global Urban Impact of Bubonic Plague, 1894–1901* (New York: New York University Press, 2007).

95. On theories of the rat's role in plague, see chapter 3.

96. Adrien Proust, *La défense de l'Europe contre la peste et la conférence de Venise de 1897* (Paris: Masson, 1897), 403.

97. Ibid., 330.

98. David S. Barnes, "'Until cleansed and purified': Landscapes of health in the interpermeable world," *Change Over Time* 6, no. 2 (November 2016), 142.

99. Ibid.

100. David S. Barnes, "Cargo, 'infection,' and the logic of quarantine in the nineteenth century," *Bulletin of the History of Medicine* 88, no. 1 (2014): 75–101. For an overview of the traffic of goods in the second half of the nineteenth century, see the excellent research results of the Trading Consequences Project: http://trading consequences.blogs.edina.ac.uk/.

101. Owsei Temkin, "An historical analysis of the concept of infection," *Studies in Intellectual History* (1953): 123–47.

102. Barnes, "Cargo, 'infection,' and the logic of quarantine in the nineteenth century," 94.

103. Ibid., 96.

104. Ibid.; J. Blancou, "History of disinfection from early times until the end of the 18th century," *Revue scientifique et technique* (International Office of Epizootics) 14, no. 1 (March 1995): 21–39; Rebecca Whyte, "Disinfection in the laboratory: Theory and practice in disinfection policy in late C19th and Early C20th England," *Endeavour* 39, no. 1 (March 2015): 35–43.

105. William Coleman, *Yellow Fever in the North: The Methods of Early Epidemiology* (Madison: University of Wisconsin Press, 1987).

106. J. Andrew Mendelsohn, "'Like all that lives': Biology, medicine and bacteria in the age of Pasteur and Koch," *History and Philosophy of the Life Sciences* 24, no. 1 (2002): 3–36, 5.

107. Thomas Schlich, "Asepsis and bacteriology: A realignment of surgery and laboratory science," *Medical History* 56, no. 3 (July 2012): 308–34; Lindsey Fitzharris, *The Butchering Art: Joseph Lister's Quest to Transform the Grisly World of Victorian Medicine* (London: Penguin, 2017).

108. Roberts Bartholow, *The Principles and Practice of Disinfection* (Cincinnati, OH: R. W. Carroll & co., 1867).

109. For example, "Burnett's Liquid" that sparked disinfectant industry in Victorian Britain, see David McLean, "Protecting wood and killing germs: 'Burnett's Liquid'

and the origins of the preservative and disinfectant industries in early Victorian Britain," *Business History* 52, no. 2 (April 2010): 285–305.

110. Bartholow, *The Principles and Practice of Disinfection,* 18.

111. Ibid., 48.

112. Whyte, "Disinfection in the laboratory," 35; Anne Hardy, *The Epidemic Streets: Infectious Disease and the Rise of Preventive Medicine, 1856–1900* (Wotton-under-Edge, UK: Clarendon Press, 1993).

113. Michael Worboys, *Spreading Germs: Disease Theories and Medical Practice in Britain, 1865–1900* (Cambridge: Cambridge University Press, 2000).

114. Mooney, *Intrusive Interventions*, 135.

115. Ibid., 136

116. Ibid.

117. William Henry, "Experiments on the disinfecting powers of increased temperatures, with a view to the suggestion of a substitute for quarantine," *Philosophical Magazine* 10, no. 9 (1832), 367

118. Mooney, *Intrusive Interventions,* 136.

119. William Henry, "Further experiments on the disinfecting powers of increased temperatures." *Philosophical Magazine* 11, no. 61 (1832): 22. William Henry, "Letter from Dr. Henry on a modified disinfecting apparatus," *Philosophical Magazine* 11, no. 63 (1832): 205.

120. Mooney (2015) has argued that Henry's heat-based disinfection method was to become the basis for the extensive guidelines for preventative disinfection produced by the chief medical officer of the Medical Department of England's Privy Council, John Simon, thirty years later. For the way in which clorine-based fumigation of households was used against "the horrific gorgon" of cholera in Scotland, and the controversy that ensued, see James Sanders, "Report of the effects of the acid fumigation tried in Scotland during the prevalence of the epidemic cholera and the causes which prevented it from being every where known and adopted, &c," *London Medical and Surgical* Journal 3 (1833): 395–400.

121. Mooney, *Intrusive Interventions*, 137.

122. Ibid., 139.

123. On how "disease surveillance generated its own forms of portability" (ambulances, disinfectors, hand held sprey pumps, etc.), see ibid, 9.

124. Ibid., 140.

125. Ibid.

126. See ibid., 141, for discussion of these.

127. Ibid.

Chapter 2

1. For a discussion of New Orleans yellow fever history, see Margaret Humphreys, *Yellow Fever and the South* (Baltimore, MD: Johns Hopkins University Press, 1999), 5.

2. Kathryn Olivarius, "Immunity, capital, and power in antebellum New Orleans," *American Historical Review* 124, no. 2 (2019): 425–455.

3. Erwin H. Ackerknecht, "Anticontagionism between 1821 and 1867," *Bulletin of the History of Medicine* 22 (1948): 562–93.

4. Louisiana Board of Health, *The Louisiana State Board of Health, Its History and Work, with a Brief Review of Health Legislation and Maritime Quarantine in Louisiana* (St. Louis, MO, 1904).

5. Ibid., 2.

6. Humphreys, *Yellow Fever and the South*, 45.

7. Ibid., 6.

8. Ackerknecht, "Anticontagionism between 1821 and 1867," 566.

9. Humphreys, *Yellow Fever and the South*, 19.

10. Ibid.

11. Ibid., 51ff.

12. Additionally, the yellow fever epidemic had been accelerated by a parallel epidemic of cholera. See Erasmus Darwin Fenner, *History of the Epidemic Yellow Fever at New Orleans, La., in 1853* (New York: Hall, Clayton, 1854).

13. Gordon Gillson, *Louisiana State Board of Health: The Formative Years* (Baton Rouge: Louisiana State Board of Health, 1976), 58.

14. Ibid.

15. Ibid., xvi.

16. Ibid., 61.

17. Ibid., 100.

18. On theories about self-generation on board of ships, see David S. Barnes, "Cargo, 'infection,' and the logic of quarantine in the nineteenth century," *Bulletin of the History of Medicine* 88: 1 (2014): 75–101.

19. Gordon Gillson, *Louisiana State Board of Health. The Progressive Years* (Baton Rouge: Louisiana State Board of Health, 1976), 161.

20. Humphreys, *Yellow Fever and the South*, 54.

21. Ibid., 27; Gillson, *Louisiana State Board of Health. The Formative Years*, 100–103.

22. Ibid., 162.

23. Kellogg, cited in ibid., 161.

24. Humphreys, *Yellow Fever and the South*.

25. Ibid., 14.

26. Charles Brainon White, *Disinfection in Yellow Fever, as Practised at New Orleans, in the Years 1870 to 1875 Inclusive* (New Orleans, LA: J. W. Madden, 1876), 4.

27. Ibid., 6.

28. A. W. Perry, "Effectual external regulations without delay to commerce," *Public Health Papers and Reports* 1 (1873): 437–440, 439.

29. White, *Disinfection in Yellow Fever*, 7.

30. Ibid., 6.

31. Ibid.

32. Louisiana Board of Health, *The Louisiana State Board of Health*, 7.

33. Freeport Sulphur Company, *4,000 years of Yellow Magic; The Story of a Basic Element, Widely Used, Little Understood, Which for 40 Centuries Has Been Shaping the History of the World* (Freeport, TX; Port Sulphur, LA: n.p., 1900).

34. Ibid.

35. Perry, "Effectual external regulations without delay to commerce," 439.

36. Ibid., 440.

37. Ibid. points to the work of the French chemist Bertholet, who had proven the equal distribution of carbolic acid in hydrogen in experiments in France just a few years earlier.

38. White, in Louisiana Board of Health, *The Louisiana State Board of Health*, 8.

39. The Matas Medical Library, Tulane University, "Louisiana Board of Health. *Annual Reports of the Louisiana State Board of Health*. New Orleans, 1875," 31.

40. Quoted in Louisiana Board of Health, *The Louisiana State Board of Health*, 9.

41. Gillson, *Louisiana State Board of Health: The Formative Years*, 167.

42. White, *Disinfection in Yellow Fever*, 14.

43. American Public Health Association, Committee on Disinfectants and Royal College of Physicians of Edinburgh, *Report of the Committee on Disinfectants of the American Public Health Association* (Baltimore: Printed for the Committee, 1885).

44. Joseph Lister, "On the antiseptic principle in the practice of surgery," *The Lancet* 2 (2299) (September 21, 1867): 353–356.

45. John Dougall, "Disinfection by acids," *British Medical Journal* 2 (984) (November 8, 1879): 726–728.

46. American Public Health Association, *Report of the Committee on Disinfectants*, 40.

47. Émile Arthur Vallin, *Traité des désinfectants et de la désinfection* (Paris: G. Masson, 1882).

48. Ibid., 487.

49. American Public Health Association, *Report of the Committee on Disinfectants*.

50. Ibid., 88.

51. Howard Markel, *Quarantine!: East European Jewish Immigrants and the New York City Epidemics of 1892* (Baltimore, MD: Johns Hopkins University Press, 1999)

52. American Public Health Association, *Report of the Committee on Disinfectants*, 91–92.

53. Robert Koch, "Ueber Desinfektion," *Mittheilungen aus dem Kaiserlichen Gesundheitsamte* 1 (1881): 287–338.

54. René Dubos, *Louis Pasteur, Free Lance of Science* (Boston: Little, Brown and Company, 1950), 139.

55. Humphreys, *Yellow Fever and the South*, 42; Louisiana Board of Health, *The Louisiana State Board of Health*, 16.

56. Joseph Holt, *Quarantine and Commerce, Their Antagonism Destructive to the Prosperity of City and State* (New Orleans: Graham and Sons, 1884), 5.

57. Humphreys, *Yellow Fever and the South*, 10ff.

58. Holt, *Quarantine and Commerce*, 1.

59. The Matas Medical Library, Tulane University, "Louisiana Board of Health. Annual Reports of the Louisiana State Board of Health. New Orleans, 1884–1885," 11.

60. Gillson, *Louisiana State Board of Health. The Progressive Years*, 11.

61. Joseph Holt, *An Epitomized Review of the Principles and Practice of Maritime Sanitation* (New Orleans: Graham, 1892).

62. Ibid., 4

63. Kennedy, cited in Gillson, *Louisiana State Board of Health. The Formative Years*, 23.

64. The Matas Medical Library, Tulane University, "Louisiana Board of Health. *Annual Reports of the Louisiana State Board of Health*. New Orleans, 1888–1889," 61.

65. Gillson, *Louisiana State Board of Health. The Formative Years*, 21.

66. Olliphant was a physician from Louisiana who had previously served as vice president of the board under Holt. He also spent a year overseeing the sanitation procedures on on the Rigolet quarantine station south of New Orleans. See *The Times Democrat*, April 11, 1890, 11.

67. "Board of Health. An Address to the Public as to Its Policy and Plans," *The Daily Picayune* (New Orleans, LA), April 23 1890, 6.

68. [No title], *St. Landry Democrat* (Opelousas, LA) 1878–1894, April 19, 1890, 1.

69. Anonymous, "Board of Health. An Address to the Public as to Its Policy and Plans," *The Daily Picayune* (New Orleans, LA), April 23, 1890, 6.

70. The Matas Medical Library, Tulane University, "Louisiana Board of Health. *Annual Reports of the Louisiana State Board of Health*. New Orleans, 1890–1891," 17.

71. Anonymous, "Trip To Quarantine. Official Visit of Inspection by the Board of Health and Legislative Committees," *The Daily Picayune* (New Orleans, LA), June 2, 1890, 3.

72. The Matas Medical Library, Tulane University, "Louisiana Board of Health. *Annual Reports of the Louisiana State Board of Health*. New Orleans, 1890–1891," 17.

73. Ibid.

74. S. R. Olliphant and T. A. Clayton, "Fumigation Apparatus," Patent US490981, filed January 25, 1892, granted January 31, 1893.

75. Ibid.

76. William Haynes, *Brimstone, the Stone That Burns: The Story of the Frasch Sulphur Industry* (Princeton, NJ: Van Nostrand, 1959), 36.

77. Louisiana Board of Health, *The System of Maritime Sanitation Inaugurated and Brought to Its Present State of Perfection by the Louisiana State Board of Health* (Chicago, 1893), 1.

78. Ibid.

79. Ibid., 11.

80. On the history of the Alliance, see Donna A. Barnes, *The Louisiana Populist Movement, 1881–1900* (Baton Rouge: Louisiana State University Press, 2011).

81. A first entry illuminating the life of Thomas A. Clayton is to be found in the *Biographical and Historical Memoirs of Louisiana* (Chicago: The Goodspeed Publishing Company, 1892), 351.

82. See *St. Landry Democrat,* April 19, 1890, 1.

83. See *The Times-Democrat* (New Orleans), May 17, 1891, 3; *The Times-Democrat* (New Orleans), July 22, 1891, 3; and *St. Landry Democrat,* June 6, 1891, 4.

84. See *The Daily Picayune,* July 24, 1894, 11.

85. New Orleans, Louisiana, City Directory, 1895, 215.

86. Anonymous, "A Case in Point," *The Washington Post* (1877–1922), December 17, 1897, 6.

87. Yet, although the disease returned in the 1890s, hardly anyone questioned the validity and functionality of the fumigation device and the system of maritime sanitation that had been developed around it.

88. Thomas A. Clayton, "Method and Apparatus for Fumigating and Extinguishing Fires in Closed Compartments," Patent US633807, filed June 7, 1899, and issued September 26, 1899.

89. A first detailed description of the capacities of the Clayton system and its advantages over other systems is given in W. J. Simpson, *A Treatise on Plague; Dealing with the Historical, Epidemiological, Clinical, Therapeutic and Preventive Aspects of the Disease* (Cambridge: Cambridge University Press, 1905), 360ff. Another description of the apparatus can be found in the papers of the National Archives (UK) See National Archives (UK), MH 19/274, "Plague Destruction of Rats on Ship; Plague Precautions. Destruction of Rats on Ships, May 24, 1900."

90. Thomas A. Clayton, Method and Apparatus for Fumigating and Extinguishing Fires in Closed Compartments. *Patent US633807.* Filed June 7, 1899, granted September 26, 1899.

Chapter 3

1. M. Alper Yalçinkaya, *Learned Patriots: Debating Science, State, and Society in the Nineteenth-Century Ottoman Empire* (Chicago: The University of Chicago Press, 2015); Miri Shefer-Mossensohn, *Science among the Ottomans: The Cultural Creation and Exchange of Knowledge* (Austin: University of Texas Press, 2015); Amit Bein, "The Istanbul earthquake of 1894 and science in the late Ottoman Empire," *Middle Eastern Studies* 44, no. 6 (2008): 909–924.

2. Zeynep Devrim Gürsel, "Thinking with X-rays: Investigating the politics of visibility through the Ottoman Sultan Abdülhamid's photography collection," *Visual Anthropology* 29, no. 3, Special Issue: Visual Revolutions in the Middle East (2016): 229–242.

3. Richard Evans, *Death in Hamburg: Society and Politics in the Cholera Years, 1830–1910* (Wotton-under-Edge, UK: Clarendon Press, 1987).

4. Neil Pemberton, "The rat-catcher's prank: Interspecies cunningness and scavenging in Henry Mayhew's London," *Journal of Victorian Culture* 19, no. 4 (2014): 520–535, 526.

5. Henry Mayhew, *London Labour and the London Poor* (London: Dover, 1860–1861).

6. Pemberton, "The rat-catcher's prank," 523.

7. Ibid.

8. Ibid., 532.

9. Ibid., 533. Rodwell was the author of a popular treatise: James Rodwell, *The Rat: Its History and Destructive Character* (London: Routledge and Co., 1858).

10. C. Renny, *Medical Report on the Mahamurree in Gurhwal in 1849–50* (Agra, India: Secundra Orphan Press, 1851).

11. Emile Rocher, *La province Chinoise du Yunnan* (Paris: Lerous, 1879); Patrick Manson, "Dr. Manson's report on the health of Amoy for the half-year ended 31st March 1878," *Customs Gazette, Medical Reports* 2 (January–March 1878): 25–27.

12. *Hong Kong Government Gazette*, GA 1895 no.146; Medical Report on the Epidemic of Bubonic Plague in 1894 (incorporating J. A. Lowson, "The Epidemic of Bubonic Plague in Hong Kong, 1894," April 13, 1895): 369–422; Alexandre Yersin, "La peste bubonique à Hong Kong," *Annales de l'Institut Pasteur* 8 (1894): 662–667.

13. M. Simond, "Paul-Louis Simond and his discovery of plague transmission by rat fleas: a centenary," *Journal of the Royal Society of Medicine* 91, no. 2 (February 1998): 101–104.

14. A. W. Bacot and C. J. Martin, "Observations on the mechanism of the transmission of plague by fleas," *Journal of Hygiene*, Plague Suppl. 3 (1914): 423–39.

15. See, for example, Nicholas Evans, "'Blaming the rat? Accounting for plague in colonial Indian medicine," *Medicine, Anthropology, Theory* 5, no. 3 (2018), doi.org/10.17157/mat.5.3.371.

16. On cholera and the Ottoman Empire, see G. Sariyildiz and O. D. Macar, "Cholera, pilgrimage, and international politics of sanitation: The quarantine station on the island of Kamaran," in *Plague and Contagion in the Islamic Mediterranean*, ed. Nukhet Varlik (Croydon, UK: ARC Humanities Press, 2017), 243–274; S. Erer and A.

D. Erdemir, "Preventive measures and treatments for cholera in the 19th century in Ottoman archive documents," *Vesalius* 16, no. 1 (June 2010): 41–48; Saurab Mishra, *Pilgrimage, Politics, and Pestilence: The Haj from the Indian Subcontinent* (Oxford: Oxford University Press, 2011). Plague had been absent from the Ottoman capital, Istanbul, since the mid 1830s; D. Panzac, *La peste dans l'Empire ottoman: 1700–1850* (Leuven: Editions Peeters, 1985).

17. Yıldırım specified that, in response to plague outbreaks in Istanbul and Anatolia, the particular "directive required for ships from the Ottoman State as well as foreign vessels that any captain coming from the Mediterranean up the Dardanelles had to show the passage certificate and passenger list to an official who approached the ship in a rowboat; if the ship was coming from an uncontaminated place it would not be stopped. If they were coming from an infected area they would be quarantined." Nuran Yıldırım, *A History of Healthcare in Istanbul: Health Organizations—Epidemics, Infections and Disease Control Preventive Health Institutions—Hospitals—Medical Education* (Istanbul: Ajansfa, 2010), 23. For a general history of quarantine in the Ottoman Empire, see D. Panzac, *Quarantaines et lazarets: L'Europe et la peste d'Orient (XVII°-XX° siècle)* (Aix-en-Provence, France: Edisud, 1986).

18. On the varying titles and the history of the establishment of the Council, see Nuran Yıldırım and Hakan Ertin, "European physicians/specialists during the cholera epidemic in Istanbul 1893–1895 and their contributions to the modernization of healthcare in the Ottoman State," in *Health Culture and the Human Body: Epidemiology, Ethics and History of Medicine; Perspectives from Turkey and Central Europe*, eds. Ilhan Likilic, Hakan Ertin, Rainer Bromer, and Hajo Zeeb (Istanbul: BETIM Center Press, 2014), 189–215.

19. A. Arslan and H. A. Polat, "Travel from Europe to Istanbul in the 19th century and the quarantine of Çanakkale," *Journal of Transport & Health* (2017): 10–17, 16. The number of foreign delegates would later be ammended in order to include Germany, the USA, and other international powers (Yıldırım 2010). International presence and influence was evident not only in the Council but also across the Ottoman Empire's emerging medical system. For discussion see: Marcel Chahrour, "'A civilizing mission'? Austrian medicine and the reform of medical structures in the Ottoman Empire, 1838–1850," *Studies in History and Philosophy of Biological and Biomedical Sciences* 38, no. 4 (December 2007): 687–705; Ceren Gülser İlikan Rasimoğlu, "The foundation of a professional group: Physicians in the nineteenth century modernizing century modernizing Ottoman Empire (1839–1908)" (Ph.D. Thesis, Boğaziçi University, 2012); Anne-Marie Moulin, "Initiating global health at the time of the Crimean war 1853–1856 and the projects of sanitary reform of the Ottoman Empire," *History of Medicine* 1 (2014): 61–79; N. Sari, "Turkey and its international relations in the history of medicine," *Vesalius* VII, no. 2 (2001): 86–93; Nermin Ersoy, Yuksel Gungor, and Aslihan Akpinar, "International Sanitary Conferences

from the Ottoman perspective (1851–1938)," *Hygiea Internationalis: An Interdisciplinary Journal for the History of Public Health* 10, no. 1 (2011): 53–79.

20. Yıldırım, *A History of Healthcare in Istanbul*, 24.

21. Ibid., 27. See 27–28 on how this system of reporting worked.

22. According to Ridvan Pasha (Istanbul's mayor) 60 percent of the quarantine organization was composed by Greeks. See Yıldırım, *A History of Healthcare in Istanbul*, 31.

23. Sylvia Chiffoleau, "Les pèlerins de La Mecque, les germes et la communauté international," *Médecine/sciences (Paris)* 27, no. 12 (December 2011): 1121–1126; Saurab Mishra, "Incarceration and resistance in a Red Sea lazaretto, 1880–1930," in *Quarantine: Local and Global Histories*, ed. Alison Bashford (London: Palgrave Macmillan, 2016), 54–65; Birsen Bulmus, *Plague, Quarantines and Geopolitics in the Ottoman Empire* (Edinburgh: Edinburgh University Press, 2012); Michael Christopher Low, *The Mechanics of Mecca: The Technopolitics of the Late Ottoman Hijaz and the Colonial Hajj* (Ph.D. Thesis, Columbia University, 2015).

24. C. Stekoulis, "Bulletin épidémiologique," *Gazette médicale d'Orient* 42, no. 11 (July 31, 1897): 169–171. See also Cozzonis Effendi, *Rapport sur la manifestation pestilentielle à Djeddah en 1898, suivi d'une esquisse sur les conditions generales de la dite ville* (Istanbul: Impr. Osmanié, 1898); Frédéric Borel, *Étude d'hygiène internationale. Choléra et peste dans le pèlerinage musulman, 1860–1903* (Paris: Masson, 1904).

25. Nükhet Varlık, *Plague and Empire in the Early Modern Mediterranean World: The Ottoman Experience, 1347–1600* (Cambridge: Cambridge University Press, 2015), 47. Varlik mentions that, although sources mention the use of rat poison (mainly arsenic-based), "there do not seem to be professional rat catchers in the early modern Ottoman Empire organized in a guild, wither because this was a service performed by another group of professionals or more likely because this was something done by individuals" (2015, 26).

26. Nicholas Taptas, "Peste bubonique. Étude clinique et bactériologique," *Gazette médicale d'Orient* 42, no. 13 (August 31, 1897): 210–214, 212. The particular doctor should not be confused with the chief doctor of Istanbul's Eftal Hospital and later the private doctor of the Turkish Prime Minister Ismet İnönü. On Taptas's other medical contributions, see Alexander Karatzanis, Constantinos Trompoukis, Ioannis Vlastos, and George Velegrakis, "On the history of modern tonsillectomy: the contribution of Nikolaos Taptas," *European Archives of Oto-Rhino-Laryngology* 268 (1687) (November 2011). https://link.springer.com/article/10.1007/s00405-011-1751-3.

27. Nicholas Taptas, "Peste bubonique. Étude clinique et bactériologique," *Gazette médicale d'Orient* 42: 15 (September 15, 1897): 227–230. On the development of this theory in the context of Hong Kong, see Staff-Surgeon Wilm, *A Report on the*

Epidemic of Bubonic Plague at Hongkong in the Year 1896 (Hong Kong: n.p., 1897). On the question of plague and the infectious corpse, see Christos Lynteris and Nicholas H. Evans (eds.), *Histories of Post-Mortem Contagion: Infectious Corpses and Contested Burials* (London: Palgrave Macmillan, 2017).

28. This was seconded by Dr. Delacour in a paper on plague originally published in the *Journal de la Chambre de Commerce française* and reprinted as Delacour, "La peste," *Revue médico-pharmaceutique* 11, no. 8 (August 15, 1898): 97–99.

29. Taptas, "Peste bubonique," 230.

30. "Correspondence Re Improvement of Tai-Ping Shan, enclosed in Robinson to Ripon, 30 August 1894," Colonial Office. Original Correspondence: Hong Kong 1841–1951; Series 129/263, CO 17303.

31. On soil and plague-related research and policy in India and Hong Kong, see Christos Lynteris, "'A Suitable soil': Plague's breeding grounds at the dawn of the third pandemic," *Medical History* 61, no. 3 (June 2017): 343–357.

32. Anonymous, "La peste," *Gazette médicale d'Orient* 44, nos. 19 and 20 (November 30 and December 15, 1899): 266–267; Anonymous, "Bulletin. La peste bubonique," *Revue médico-pharmaceutique* 11, no. 8 (August 15, 1898): 101–102. For an account of the 1899 plague in Alexandria, see Aristide Valassopoulo, *La peste d'Alexandrie en 1899 au point de vue clinique, épidémiologique, etc.* (Paris: A. Maloine, 1901). For the history of plague in nineteenth-century Egypt, see LaVerne Kuhnke, *Lives at Risk: Public Health in Nineteenth-Century Egypt* (Berkeley, CA: University of California Press, 1990); Echenberg, Myron, *Plague Ports: The Global Urban Impact of Bubonic Plague, 1894–1901* (New York: New York University Press, 2007).

33. C. Stekoulis, "Bulletin épidémiologique," *Gazette médicale d'Orient* 43: 24 (February 15, 1899): 353–354.

34. Anonymous, "La peste," 266. The Imperial Society of Health (Cemiyet-I Tibbiye-I Sahane) was formed in 1856 by foreign doctors who remained in the Ottoman capital after the end of the Crimean War (during which they had been employed by the Ottoman army) with the aim of "investigat[ing] the epidemics and prepar[ing] reports which were then presented to the relevant authorities"; Yıldırım and Ertin, "European physicians/specialists during the cholera epidemic in Istanbul 1893–1895," 189.

35. Stekoulis, "Bulletin épidémiologique."

36. C. Stekoulis, "La peste—La cholera," *Gazette médicale d'Orient* 44, nos. 23 and 24 (January 30 and February 15, 1900): 307–308.

37. Ibid., 307

38. In ibid., 308.

39. Michele Nicolas, "Pierre Apéry et ses publications scientifiques," *Revue d'histoire de la pharmacie* 94, no. 350 (2006): 237–247; Halil Tekiner and Afife Mat, "Les pharmacopées turques de langue française," *Revue d'histoire de la pharmacie* 57, no. 361 (2009): 17–22.

40. Alfred Swaine Taylor, "An account of the Grotta del Cane; with remarks on suffocation by carbonic acid," *London Medical and Physical Journal* NS 73 (October 1832): 278–285.

41. Pierre Apéry, "Moyen de destruction des rats à bord des bateaux surtout en temps d'épidémie de peste," *Revue médico-pharmaceutique* 12, no. 10 (October 15, 1899): 137–138.

42. Société Imperiale de Médecine, "Séance du 4 mai 1900; L'anhydride carbonique comme moyen de destruction des rats dans les cales des bateaux," *Gazette médicale d'Orient* 45, no. 7 (May 30, 1900): 108–113.

43. Ibid., 109.

44. On this premechanized method of producing SO_2, see Pierre Apéry, "Note sur l'emploi de l'anhydride carbonique pour la destruction des rats dans les cales des navires," *Revue médico-pharmaceutique* 15, no. 1 (January 1, 1902): 1–3.

45. Société Imperiale de Médecine, "Séance du 4 mai 1900; L'anhydride carbonique comme moyen de destruction des rats dans les cales des bateaux," *Gazette médicale d'Orient* 45, no 7 (May 30, 1900), 112.

46. Ibid., 113.

47. Société Imperiale de Médecine, "Séance du 18 mai 1900; L'anhydride carbonique comme moyen de destruction des rats dans les cales des bateaux, surtout en temps d'épidémie de peste," *Gazette médicale d'Orient* 45, no. 9 (June 30, 1900): 134–140.

48. Ibid., 138.

49. Pierre Apéry, "Lettre ouverte, Reponse à M. Desguin, L'anhydride carbonique peut-il debarasser des rats les cales des bateaux?," *Revue médico-pharmaceutique* 13, no. 8 (April 15, 1900): 86–89. Invented by Petrus Jacobus Kipp (c. 1844), the Kipp apparatus would need to be massively scaled up for such an operation.

50. Société Imperiale de Médecine, "Séance du 18 mai 1900," 138–139.

51. Ibid.

52. Pierre Apéry, "Lettre ouverte, Reponse à M. Desguin."

53. Ibid., 88.

54. Anonymous, "Destruction des rats à bord des navires," *Revue médico-pharmaceutique* 13, no. 23 (December 1, 1900): 266.

55. Anonymous, "Memoire sur l'influence des rats dans la propagation de la peste," *Revue médico-pharmaceutique* 13, no. 22 (November 15, 1900): 253–257.

56. Andreas David Mordtmann, "Bulletin sanitaire," *Gazette médicale d'Orient* 45, no.7 (May 30, 1900): 102.

57. Istanbul had been free from plague since 1838; Yıldırım, *A History of Healthcare in Istanbul*, 28.

58. Andreas David Mordtmann (1837–1912), not to be confused his son, the Orientalist Johannes Heinrich Mordtmann, was the German delegate to the Quarantine Council and previously the Consul of the Hanseatic League to the Sublime Porte. On his work on cholera, see Yıldırım and Ertin, "European physicians/specialists during the cholera epidemic in Istanbul 1893–1895."

59. Andreas David Mordtmann, "Bulletin épidémiologique," *Gazette médicale d'Orient* 45, no. 14 (September 15, 1900): 267, 268.

60. C. Stekoulis, "Bulletin épidémiologique," *Gazette médicale d'Orient* 45, no. 17 (October 31, 1900): 331–332, 331. On quarantine being an affliction more severe than plague itself, see C. Stekoulis, "Bulletin épidémiologique," *Gazette médicale d'Orient* 46, no. 13 (September 1, 1901): 738–739. The antiquarantine sentiment was particularly fuelled by the proquarantine attitude of Bulgarian sanitary authorities, described by Mordtmann as a "sinister plan" involving "subjecting human beings" to unacceptable conditions of hygienic detainment; Andreas David Mordtmann, "Bulletin sanitaire," *Gazette médicale d'Orient* 45, no. 23 (January, 31 1901): 459–460, 459.

61. C. Stekoulis, "Bulletin épidémiologique," *Gazette médicale d'Orient* 46, no. 6 (May 15, 1901): 627; on Istanbul cases and their benign nature, see C. Stekoulis, "Bulletin épidémiologique," *Gazette médicale d'Orient* 46, no. 12 (August 15, 1901): 721–723.

62. C. Stekoulis, "Bulletin épidémiologique," *Gazette médicale d'Orient* 46, no. 13 (September 1, 1901): 738–739; Spiridion C. Zavitziano, "Bulletin épidémiologique," *Gazette médicale d'Orient* 46, no. 14 (September 15, 1901): 754.

63. C. Kelaiditis, "Opinion [sur] S. Balis, 'Contribution à l'étude des mesures sanitaires contre la propagation de la peste,'" *Revue medico-pharmaceutique* 15, no. 1 (January, 1 1902): 18.

64. Andreas David Mordtmann, "Bulletin épidémiologique," *Gazette médicale d'Orient* 46, no. 18 (November, 15 1901): 818.

65. Société Imperiale de Médecine, "Séance du 29 novembre 1901: Les quarantaines," *Gazette médicale d'Orient* 46, no. 22 (January 15, 1902): 887.

66. Ibid., 888. For a proquarantine response, see S. Serpossian, "La question des quarantaines, au sein de la Société Imperiale de Médecine," *Revue médico-pharmaceutique* 15, no. 6 (March 15, 1902): 67; S. Serpossian, "La question des quarantaines, au sein de la Societe Imp. de Medicine," *Revue médico-pharmaceutique* 15, no. 7 (April 1, 1902): 76.

67. Pierre Apéry, "Bulletin épidémiologique," *Revue médico-pharmaceutique* 14, no. 16 (August 15, 1901): 181.

68. Pierre Apéry, "Bulletin épidémiologique. L'utilité des quarantaines," *Revue médico-pharmaceutique* 14, no. 21 (November 1, 1901): 241.

69. Ibid.

70. Ibid.

71. Société Imperiale de Médecine, "Séance du 13 decembre 1901, Les quarantaines," *Gazette médicale d'Orient* 46, no. 23 (February 1, 1902): 905.

72. Ibid., 908; It is not hard to imagine the frustration of Apéry's colleagues with this position, especially as only a few days earlier he himself had published an editorial in the *Revue médico-pharmaceutique* titled "Quarantined suppressed"; Pierre Apéry, "Les quarantaines supprimées," *Revue médico-pharmaceutique* 14, no. 23 (December 1, 1901): 265. Indeed, Apéry's editorial was read by Stchepotiew as suggesting that his method would lead, sooner or later, to the suppression of quarantines; Société Imperiale de Médecine, "Séance du 20 decembre 1901," *Gazette médicale d'Orient* 46 (24) (February 15, 1902): 915–917. However, the title, as Apéry explained, did not express his personal position but was the verbatim title of the French article from *Le Petit Journal* (December 7, 1901) that the editorial was referring to.

73. Pierre Apéry, "Bulletin épidémiologique. L'utilité des quarantaines," *Revue médico-pharmaceutique* 14, no. 21 (November 1, 1901): 241.

74. Ibid.

75. Ch. Bonkowski, "De l'emploi de l'anhyhride carbonique pour la destruction des rats dans les cales des navires," *Revue médico-pharmaceutique* 14, no. 23 (December 1, 1901): 264–267.

76. Charles Zitterer, "Rapport sur les experiences de dératisation des cales des navires par le gaz acide carbonique présenté dans le Conseil Superieur de Santé (séance du 8/21 janvier 1902)," *Revue médico-pharmaceutique* 15, no. 1 (January 1, 1902): 14–16.

77. Key bibliography on this includes: Suraiya Faroqhi (ed.), *Animals and People in the Ottoman Empire* (Istanbul: Eren, 2010); Alpaslan Demir, "Mice problems in the Ottoman Empire and mice invasion in Tirhala in 1866," *IBAC* 2 (2012):

645–661; Alan Mikhail, *The Animal in Ottoman Egypt* (Oxford: Oxford University Press, 2014).

78. Robert Koch [Interview]. *Hamburger Freie Presse*, November 26, 1892.

79. Richard Evans, *Death in Hamburg: Society and Politics in the Cholera Years, 1830–1910* (Wotton-under-Edge, UK: Clarendon Press, 1987).

80. Bernhard Nocht, Emil von Behring, Joseph Brix, E. Pfuhl, and the Royal College of Physicians of Edinburgh, *Die Bekämpfung der Infectionskrankheiten : Hygienischer Theil* (Leipzig, Germany: G. Thieme, 1894), 472.

81. Bernhard Nocht, "Ueber die Abwehr der Pest," *Archiv fuer Schiffs und Tropen Hygiene* 1, no. 2 (1897): 92.

82. Ibid., 97.

83. An exception to this rule were the vessels of the Hamburg-American Line, which were required to place rat poison upon arrival but not to undergo fumigation each time, but only upon the instruction of the port physician.

84. C. Smith, "Germany. Method of killing rats at Hamburg," *Public Health Reports (1896–1970)* 15, no. 17 (April 27, 1900): 1013.

85. G. Mazaraky, *Le rôle des rats dans la propagation de la peste* (Thèse de méd., Fac. de Médecine de Paris; Paris: Vigot Frères, 1901).

86. Bernhard Fleischer, "A century of research in tropical medicine in Hamburg: the early history and present state of the Bernhard Nocht Institute," *Tropical Medicine and International Health* 5, no. 10 (2000): 747–751.

87. The damaging capacity was according to Nocht solely ascribed to Pictolin containing sulphur. Nocht was convinced that therefore the Clayton gas, which was equally based on sulphur, was also to be ruled out as a means of establishing a reliable fumigation system in the port of Hamburg.

88. Ministère des Affaires étrangère, *Conférence sanitaire internationale de Paris: 10 octobre-3 décembre 1903, procès-verbaux* (Paris: Impr. Nationale, 1904), 369.

89. Bernhard Nocht and G. Giemsa, "Über die Vernichtung von Ratten an Bord von Schiffen. Als Massregel gegen die Einschleppung der Pest," *Arbeiten aus Dem Kaiserlichen Gesundheitsamte* XX, no. 1 (1904), 102.

90. Ibid., 107.

91. Ibid., 113.

92. Anonymous, "Method of fumigation of vessels at Hamburg," *United States Naval Medical Bulletin* 3 (1909): 368–370.

Chapter 4

1. Advertisement in *Industrial Management* 21 (1901): 998ff.

2. Nayan Shah, *Contagious Divides: Epidemics and Race in San Francisco's China-town* (Berkeley. CA: University of California Press, 2001), 60. See also Guenter B. Risse, *Driven by Fear: Epidemics and Isolation in San Francisco's House of Pestilence* (Champaign, IL: University of Illinois Press, 2015). Markel writes on the extensive introduction of fumigation in late nineteenth-century New York: Howard Markel, *Quarantine!: East European Jewish Immigrants and the New York City Epidemics of 1892* (Baltimore, MD: Johns Hopkins University Press, 1999), 50. Also see *The St. Paul Globe* (Minnesota), February 7, 1904, 21.

3. Anonymous, *New York Times*, October 30 1910, 17

4. Krista Maglen, *The English System. Quarantine, Immigration and the Making of a Port Sanitary Zone* (Manchester, UK: Manchester University Press, 2014), 2.

5. Ibid., 8.

6. Ibid., 40.

7. Ibid., 40. Neither Maglen nor the corresponding "Report of the Medical Officer of Health for Port of London" specify how fumigation was employed or what sub-stance was used.

8. National Archives (UK), MH 19/274, "Plague Destruction of Rats on Ship; Plague Precautions. Destruction of Rats on Ships, 24 May 1900."

9. Ibid.

10. In fact, experiments regarding the fire-extinguishing properties of the Clayton machine had already taken place on January 18 of the same year at the Northumber-land forge of the North-Eastern Marine Engineering Company in the presence of the president of the company, T. A. Clayton himself.

11. Ibid.

12. Ibid.

13. Ibid.

14. Ibid.

15. Ibid.

16. Anonymous, "The plague. The destruction of rats on shipboard," *The British Medical Journal* 1, no. 2106 (May 11, 1901), 1170. At the same time, similar experi-ments or demonstrations were taking place in other parts of the world, such as South Africa's Cape Town, where the Clayton was used to fumigate the steamship

Goorkha, achieving satisfactory results as regards vermin eradication and no tarnishing of materials; National Archives (UK), MH 19/274, "Plague Destruction of Rats on Ship; Plague Precautions. Destruction of Rats on Ships, 24 May 1900," enclosed pamphlet "The Clayton Fire Extinguishing and Ventilating Company, Limited."

17. Ibid.

18. "The Clayton Fire Extinguishing and Ventilating Company, Thomas A. Evans, to the Assistant Secretary, Marine Department, Board of Trade, Whitehall, October 29, 1901." National Archives (UK), MH 19/274, "Plague Destruction of Rats on Ship; Plague Precautions. Destruction of Rats on Ships, 24 May 1900."

19. M. Simond, "Paul-Louis Simond and his discovery of plague transmission by rat fleas: a centenary." *Journal of the Royal Society of Medicine* 91, no. 2 (February 1998): 101–104.

20. Rechit Khayat, *Prophylaxie de la peste par la destruction des insectes et des rongeurs* (Paris: Jules Rousset, 1902), 7. See also Jules Bucqcoy, "Une quarantaine au Frioul (séance du 29 octobre 1901)," *Bulletin de l'Académie nationale de médecine* 3, no. XLVI (1901), 422–436.

21. On the history of the scientific cruises of the *Revue générale des sciences pures et appliquées,* see Veronica della Dora, "Making mobile knowledges: The educational cruises of the Revue générale des sciences pures et appliquées, 1897–1914," *Isis* 101, no. 3 (September 2010): 467–500.

22. The man died in quarantine.

23. The experiences of the passengers were recorded in a memoir, including photographs by the pioneer of French cinema Léon Gaumont. See Jean Bertot, *Au lazaret. Souvenirs de quarantaine* (Tours, France: Deslis frères, 1902). See also Jacques Chevallier, "Une quarantaine de peste au lazaret de Frioul en 1901," *Histoire des sciences médicales* XLIX, no. 2 (2015): 179–188.

24. The role of rats in Alexandria had been noted by local medical authorities. See Aristide Valassopoulo, *La peste d'Alexandrie en 1899 au point de vue clinique, épidémiologique, etc.* (Paris: A. Maloine, 1901).

25. Anonymous, "La peste à bord du Sénégal," *Petit Journal* (December 1, 1901): 264; Anonymous, "Le lazaret du Frioul," *L'actualité* 91 (October 20, 1901): 1. For photographic coverage, see Anonymous, no title, *La Revue hebdomadaire* 4, no. 47 (October 19, 1901): 2–4.

26. Joseph Pellisier, *La peste au Frioul, lazaret de Marseille en 1900 et 1901* (Paris: Steinheil, 1902).

27. G. Mazaraky, *Le rôle des rats dans la propagation de la peste* (Thèse de méd., Fac. de Médecine de Paris. Paris: Vigot Frères, 1901), 81–82; see also Henri Monod, "Incident

du paquebot 'Sénégal' retenu au lazaret de Frioul le 20 september 1901 comme infecté de peste," *Recueil des travaux du Comité Consultatif d'Hygiène Publique de France et des Actes Officiels de l'Administration Sanitaire* 31 (1901), 186 note 1; on a review of rat studies in light of the *Sénégal* outbreak, see Pellisier, *La peste au Frioul.*

28. Pellisier, *La peste au Frioul*, 180; on the other incidents concerning vessels such as the *Niger*, the *Laos*, and the *Simla*, see A. Proust and Paul Faivre, "Rapport général sur les maladies pestilentiels exotiques (peste, fièvre jaune, cholera) en 1901," *Recueil des Travaux du Comité Consultatif d'Hygiène Publique de France et des Actes Officiels de l'Administration Sanitaire* 31 (1901), esp. 290–332.

29. These included the instructions of August 4, 1899, which recommended the sulphurization of boats and lazarettos against rats and mice, and the instructions of October 1, 1900, which instituted the surveillance of unloading of boats deriving from plague-infected ports for sick or dead rats.

30. This did not necessarily mean the use of the Clayton machine, but referred more broadly to the use of sulphur as a fumigant aganst rats. See Henri Monod, "Incident du paquebot 'Sénégal.'"

31. In spite of Proust's self-defensive insistence that previous deratization rules sufficed, they in fact only mentioned rats tentatively. See Anonymous, "Discussions: Sur la police sanitaire maritime et le sejour du *Sénégal* au lazaret du Frioul (Séance 5 novembre 1901)," *Bulletin de l'Académie nationale de médecine* 3, no. XLVI (1901): 488–503.

32. Conseil Quarantenaire d' Égypte, "Rapport adressé à la presidence du Conseil Quarantinaire d' Égypte par M. le directeur P. I. de l'Office Quarantinaire d'Alexandrie sur la destruction des rats à bord des navires," *Recueil des Travaux du Comité Consultatif d'Hygiène Publique de France et des Actes Officiels de l'Administration Sanitaire* 31 (1901): 513–516.

33. Ibid., 514.

34. Émile Arthur Vallin, "Rapport. Les services sanitaires et le lazaret du Frioul (séance 11 mars 1902)," *Bulletin de l'Académie nationale de médecine* 3, no. XLVII (1902), 332.

35. Ibid., 333.

36. A. Proust and Paul Faivre, "Rapport sur different procédés de destruction des rats à bord des navires," *Recueil des Travaux du Comité Consultatif d'Hygiène Publique de France et des Actes Officiels de l'Administration Sanitaire* 33 (1903), 371.

37. Vallin, "Rapport. Les services sanitaires et le lazaret du Frioul," 333.

38. Émile Arthur Vallin, *Traité des désinfectants et de la désinfection* (Paris: G. Masson, 1882).

39. Vallin, "Rapport. Les services sanitaires et le lazaret du Frioul."

40. Service Sanitaire Maritime, "Sulfuration des navires.—Emploi d'appareils permettant d'effectuer cette sulfuration avant déchargement, circulaire du president du conseil, ministre de l'interieur et des cultes, du 20 juillet 1903, aux directeurs de la santé," *Recueil des Travaux du Comité Consultatif d'Hygiène Publique de France et des Actes Officiels de l'Administration Sanitaire* 33 (1903), 103.

41. Adrien Loir, a nephew of Louis Pasteur, was the founder of the Pasteur Institute of Tunis and an internationally reknowened bacteriologist.

42. Wellcome Library, Reference number b19884217, Port of London Sanitary Committee, "Annual Report of the Medical Officer to 31st December 1902," 30.

43. Proust and Faivre, "Rapport general sur les maladies pestilentiels exotiques," 307.

44. Gustave Duriau, "Quelques mots sur l'appareil Clayton," *Mémoires de la Société Dunkerquoise pour l'encouragement des sciences, des lettres et des arts* 36 (1902), 403.

45. Pierre Apéry, "De l'inefficacité des systems à base d'anhydride sulfureux employés jusqu'à ce jour pour la destruction des rats dans des navires," *Revue médico-pharmaceutique* 15, no. 4 (February 15, 1902): 36–39.

46. J.-P. Langlois and A. Loir, "La destruction des rats à bord des bateaux comme mesure prophylactique contre la peste," *Revue d'hygiène et de police sanitaire* 24 (1902), 413.

47. On Rosenstiehl, see Anne-Claire Déré, "Daniel August Rosenstiehl (1839–1916): An Alsatian chemist in the synthetic dyestuffs industry," in *The Chemical Industry in Europe, 1850–1914: Industrial Growth, Pollution, and Professionalization*, eds. Ernst Homburg, Anthony S. Travis, and Harm G. Schröter (Dordrecht, Netherlands: Springer, 1998), 305–320.

48. National Archives (UK), MH 19/274, "Plague Destruction of Rats on Ship; Report on the Disinfection of Vessels by the Clayton System by Dr. A. Calmette, Director of the Pasteur Insitute of Lille, and Dr. Haufeuille Assistant at the Pasteur Insitute of Lille," translated from the *Revue d'Hygiene et de police sanitaire, Paris* XXIV, no. 10 (October 20, 1902): 22–27.

49. Ibid.

50. Proust and Faivre, "Rapport sur different procedes de destruction des rats à bord des navires," 356.

51. National Archives (UK), MH 19/274, "Plague Destruction of Rats on Ship; Report on the Disinfection of Vessels by the Clayton System," 6.

52. Proust and Faivre, "Rapport sur different procedes de destruction des rats à bord des navires," 363.

53. Ibid., 364.

54. Ibid.

55. Ibid.

56. Ibid. 379.

57. Ibid., 380.

58. ibid., 350.

59. M. E. David and G. Duriau, "État actuel de la désinfection des navires. Carbonication, sulfuration (procédé Clayton)," *Revue d'hygiène et de police sanitaire* 25 (1903), 502.

60. Sanitary Administration of the Ottoman Empire, "Circular No.180. Instructions Concerning Vessels which Have or Have Not Undergone Disinfection in View of Destroying Rats and Mice on Board" (November 20/December 2, 1901), enclosed in Consul-General Constantinople, "Extermination of Rats, July 6 1909"; NARA RG90 Central File 1897–1923 537–544 Box 065.

61. Ibid.

62. National Archives (UK), MH 19/274, "Plague Destruction of Rats on Ships, 1899–1903, Franck Clemow to de Bunsen, Constantinople, August 16, 1902."

63. National Archives (UK), MH 19/274, "Plague Destruction of Rats on Ships, 1899–1903, Communication faite en Séance du Conseil Superieur de Santé, le 16/29 Juillet 1902, par Monsieur le Délégué d'Angleterre."

64. National Archives (UK), MH 19/274, "Plague Destruction of Rats on Ships, 1899–1903, Clemow, Constantinople October 12, 1902."

65. Ibid.

66. Ibid.

67. S. Balilis, *Contribution à l'étude des mesures sanitaires contre la propagation de la peste* (Constantinople: Impr. A. Zellich fils, 1901).

68. Ibid., 4.

69. Ibid., 5.

70. Ibid., 6.

71. Ibid.

72. National Archives (UK), MH 19/274, "Plague Destruction of Rats on Ships, 1899–1903, Clemow, Constantinople October 12, 1902."

73. Ibid.

74. Such as Apéry's machine, which was already mentioned favourably in Balilis, *Contribution à l'étude des mesures sanitaires.*

Chapter 5

1. Anonymous, "The destruction of rats on ship-board," *The British Medical Journal* 1, no. 2144 (February 1, 1902): 294–295.

2. Ibid., 294.

3. National Archives (UK), MH 19/274, "Plague Destruction of Rats on Ship; W. S. Power to Sir Provis, 18 June 1903."

4. Ibid.

5. Ibid.

6. National Archives (UK), MH 19/274, "Plague Destruction of Rats on Ship, Peter Samson, Report 10/06/1903."

7. National Archives (UK), MH 19/274, "Plague Destruction of Rats on Ship, E. C. Blech to H. R. O'Connor, July 9 1903"; Dr Duca's appointment to the Quarantine Council to replace the retiring Dr. Cozonnis was the subject of considerable controversy. See Spiridion C. Zavitziano, "Turkey. Report from Constantinople," *Public Health Reports (1896–1970)* 16, no. 12 (March 22, 1901): 602–604.

8. National Archives (UK), MH 19/274, "Plague Destruction of Rats on Ship, E. C. Blech to H. R. O'Connor, July 9 1903."

9. While not clear in existing archival documents, the reason for the choice of Clemow might have been that he had both previously observed fumigation experiments in Istanbul and expressed doubts about the Clayton process (see chapters 3 and 4), which would render him a reliable observer.

10. National Archives (UK), MH 19/274, "Plague Destruction of Rats on Ship, Frank G. Clemow, Report of Experiment on the steamship 'Westmoreland' to test the efficacy of the 'Clayton' Process, July 30 1903, Handwritten note, 13 August 1903."

11. Ibid.

12. Ibid.

13. Ibid. Clemow attributed the unusual killing of the anthrax spores to them being "an old laboratory culture."

14. Ibid.

15. Ibid.

16. Ibid.

17. Frédéric Borel, *Étude d'hygiène internationale. Choléra et peste dans le pèlerinage musulman, 1860–1903* (Paris: Masson, 1904).

18. Borel ("Séance du 4 septembre") in *XXII Congrès international d'hygiène et de démographie, Copte rendus du Congrès, Tome VII Premiere Division—Hygiène, Section VI. Hygiène administrative*. Brussels: P. Weissenbruch, 1903, 129.

19. Ibid.

20. Ibid.

21. Calmette in ibid., 134.

22. Nocht in ibid.

23. Nocht in ibid., 117.

24. See chapter 3.

25. Langlois in ibid., 130.

26. Actes officiels, "Destruction des rats à bord des navires provenant de pays contaminés de peste, avant déchargement, décret du 21 septembre 1903," *Recueil des actes officiels et documents intéressant l'hygiène publique (Travaux du Conseil supérieur d'hygiène publique de France)* 33 (1903): 106–107.

27. S. W. Sturdy, "A co-ordinated whole: The life and work of John Scott Haldane" (Ph.D. Thesis, University of Edinburgh, 1987). https://www.era.lib.ed.ac.uk/handle/1842/6873.

28. Ibid., 152.

29. "Reports to the Local Government Board on the Destruction of Rats and Disinfection on Shipboard," by J. S. Haldane, M.D., F.R.S., and John Wade, D.Sc, I. "Observations of the Clayton process as used at Dunkirk," by Dr. Haldane, British Library, Indian Office, IOR/R/20/A/2596 File 118/7 Plague: Rat destruction, 2.

30. Ibid.

31. Ibid.

32. Ibid.

33. Ibid., 3.

34. Ibid.

35. Ibid.

36. Ibid.

37. Ibid.

38. Ibid.

39. "Reports to the Local Government Board on the Destruction of Rats and Disinfection on Shipboard," by J. S. Haldane, M.D., F.R.S., and John Wade, D.Sc, II. "Experiments on the Clayton process and sulphur dioxide as applied in the destruction of rats and in disinfection," by Dr. Wade, British Library, Indian Office, IOR/R/20/A/2596 File 118/7 Plague: Rat destruction, 3.

40. Ibid., 6.

41. Ibid., 6

42. Ibid., 7.

43. Ibid., 10.

44. Ibid., 13.

45. Ibid., 14.

46. Ibid., 18, emphasis in the original.

47. Ibid., 19

48. Ibid.

49. "Reports to the Local Government Board on the Destruction of Rats and Disinfection on Shipboard," by J. S. Haldane, M.D., F.R.S., and John Wade, D.Sc, III. "Discussion of the process for destruction of rats on shipboard," by Drs. Haldane and Wade, British Library, Indian Office, IOR/R/20/A/2596 File 118/7 Plague: Rat destruction, 27.

50. Ibid.

51. Ministère des Affaires étrangère, *Conférence sanitaire internationale de Paris: 10 octobre-3 décembre 1903, procès-verbaux* (Paris: Impr. Nationale, 1904), 24.

52. Ibid., 25.

53. Ibid., 24.

54. Ibid., 45.

55. Ibid., 94.

56. Ibid., 368.

57. Ibid., 369–370.

58. Ibid.

59. Ibid., 372.

60. Ibid., 374.

61. Ibid., 375.

62. Ibid., 375.

63. Anonymous, "Décret concernant la destruction des rats à bord des navires provenant de pays contaminés de peste (4 mai 1906)," *Bulletin officiel du ministère de l'intérieur* 5 (1906): 261–263.

64. Ibid., 262.

65. Ibid.

66. Ibid.

67. Ibid.

68. For an overview of sequential decrees leading to the 1904 decree, see Barthélemy and Dr. Georges Varenne, *Manuel d'hygiène navale: À l'usage des capitaines, des officiers et des élèves de la marine marchande* (Paris: Challamel, 1907).

69. See, for example, Ruth Rogaski's study of how this international antagonism played out in Chinese treaty ports like Tianjin. Ruth Rogaski, *Hygienic Modernity. Meanings of Health and Disease in Treaty-Port China* (Berkeley, CA: University of California Press, 2004).

70. On Zuschlag's rat-related research, see Emil Zuschlag, *Le rat migratoire et sa destruction rationnelle*, trans. M. Pierre Oesterby (Copenhagen, Impr. F. Bagge, 1903).

71. "US Consul General Copenhagen, January 16, 1909, Extermination of Rats in Denmark," NARA RG90 Central File 1897–1923 537–544 Box 065. The Danish state-organized war against rats may here be compared to the one in colonial Hanoi in 1902, as examined in Michael G. Vann, "Of rats, rice, and race: The great Hanoi rat massacre, an episode in French colonial history," *French Colonial History* 4 (2003): 191–203.

72. See, for example, Dorothy Worell, *The Women's Municipal League of Boston: A History of Thirty Five Years of Civic Endeavor* (Boston: Women's Municipal League of Committees Inc., 1943); Mrs. Albert T. Leatherbee and the Women's Municipal League of Boston, *Plague Conditions in Boston*, 1921. [Pamphlet]; Anonymous, "Elimination of the rat," *Boston Medical and Surgical Journal* 174, no. 2 (October 19, 1916): 576; the said article was the following M. W. Richardson, "The rat and infantile paralysis: A theory," *Boston Medical and Surgical Journal* 175 (September 21, 1916): 397–400. Similar campaigns were frequently organized around a Rat Day or

Rat Week theme across the East Coast. For examples, see NARA RG90 Central File, 1897–1923, 544 Box 066.

73. Branwyn Poleykett, "Building out the rat: Animal intimacies and prophylactic settlement in 1920s South Africa," *Engagement* (February 7, 2017). https://aesengagement.wordpress.com/2017/02/07/building-out-the-rat-animal-intimacies-and-prophylactic-ssettlement-in-1920s-south-africa/.

74. David E. Lanz, *How to Destroy Rats* (Washington, DC: Government Printing Office, 1909). For example, during the September 1915 plague epizootic in New Orleans, premises from 801 to 845 South Fulton Street were subjected to fumigation with hydrocyanic acid gas; "US Public Health Service to The Surgeon-General, Public Health and Marine-Hospital Service, Washington D.C., February 25 1921, Report on Recent Outbreak of Rodent Plague"; NARA RG90 Central File, 1897–1923, 544 Box 066.

75. Anonymous, "Ship rats and plague," *Public Health Reports (1896–1970)* 29, no. 16 (April 17, 1914): 927–928.

76. "Consul-General Constantinople, Extermination of Rats, July 6, 1909"; NARA RG90 Central File 1897–1923 537–544 Box 065; "Passed Assistant Surgeon, American Consulate Amoy to Public Health and Marine-Hospital Service" (non-dated); NARA RG90 Central File 1897–1923 537–544 Box 065

77. "Report on Extermination of Rats at Foreign Ports, Trieste, Austria, March 4, 1909"; NARA RG90 Central File 1897–1923 537–544 Box 065.

78. Jaime Larry Benchimol, *Dos micróbios aos mosquitos: Febre amarela e a revolução pasteuriana no Brasil*, (Rio de Janeiro: SciELO—Editora FIOCRUZ, 1999).

79. "Consul General Callao Peru to Assistan Secretary of State, February 4. 1909, Extermination of Rats at Foreign Seaports"; NARA RG90 Central File 1897–1923 537–544 Box 065. However, at the same time, other methods and machines for rat destruction onboard ships included not only innovations in the field of fumigation but also in other deratization technologies, such as trapping and poisoning. Examples include the invention and experimentation with the German-made Ratin, "a pasty substance inoculated with a bacillus" in India (1907) ("Calcutta US Consul-General to the Assistant Secretary of State, May 27 1909, Destruction of Rats"; NARA RG90 Central File 1897–1923 537–544 Box 065). Or indeed the application in Vladivostok of "typhus bacillus in jelly" prepared by the Central Laboratory of Milk Industry in Tomsk, which was mixed with caustic lime so as to be made into "bread balls" ("US Consul Vladivostok, March 24 1909, Extermination of Rats"; NARA RG90 Central File 1897–1923 537–544 Box 065). The result of the latter method was that having eaten the bait, rats became very thirsty and sought to drink water, which reacted with the lime in its stomach, causing death. The method proved unsatisfactory as it was applicable only in buildings, while rats did not really find the bait so attractive.

80. Sokhieng Au, *Mixed Medicines: Health and Culture in French Colonial Cambodia* (Chicago, IL: The University of Chicago Press, 2011); Christos Lynteris, "Vagabond microbes, leaky labs and epidemic mapping: Alexandre Yersin and the 1898 plague epidemic in Nha Trang," *Social History of Medicine* (2019), https://doi.org/10.1093/shm/hkz053.

81. ANOM, INDO GGI 4416, Appareil Clayton.

82. On blaming Chinese junks as spreaders of plague from China to French Indochina, see Lynteris, "Vagabond microbes, leaky labs and epidemic mapping."

83. Ibid.

84. Ibid.

Chapter 6

1. As a history pertinent to the peculiar space of urban modernity that developed in Buenos Aires at the turn of the twentieth century, this local case also requires differentiation from the health transformation affecting the nation of Argentina and very different trajectories experienced at the same time in rural communities. See María Silvia Di Liscia, "Del brazo civilizador a la defensa nacional: Políticas sanitarias, atención médica y población rural (Argentina, 1900–1930)," *Historia Caribe* 12, no. 31 (2017): 159–93.

2. Marcos Cueto and Steven Palmer, *Medicine and Public Health in Latin America* (Cambridge: Cambridge University Press, 2014), 13.

3. Diego Armus, "Disease in the historiography of modern Latin America," in *Disease in the History of Modern Latin America: From Malaria to AIDS*, ed. Diego Armus (Durham, N.C.: Duke University Press, 2003), 13; Marcos Cueto, *The Value of Health: A History of the Pan American Health Organization* (Washington, DC: Pan American Health Organization, 2007).

4. José Amador, *Medicine and Nation Building in the Americas, 1890–1940* (Nashville, TN: Vanderbilt University Press, 2015), 5.

5. J. Guilherme Lacorte, "A atuação de Oswaldo Cruz no aparecimento da peste bubônica no Brasil," *A Folha Medica* 54 (1967): 183–188.

6. Fernanda Rebelo, "Between the Carlo R. and the Orleannais: Public health and maritime prophylaxis in the description of two cases of ships transporting immigrants arriving in the Port of Rio de Janeiro, 1893–1907," *História, Ciências, Saúde-Manguinhos* 20, no. 3 (2013), 12.

7. Jaime Larry Benchimol, *Dos micróbios aos mosquitos: Febre amarela e a revolução pasteuriana no Brasil* (Rio de Janeiro: SciELO—Editora FIOCRUZ, 1999), 259ff.

8. Alejandro Kohl, *Higienismo Argentino: Historia de una utopía. La salud en el imaginario colectivo de una época* (Buenos Aires: Editorial Dunken, 2006); Kristin Ruggiero, *Modernity in the Flesh: Medicine, Law, and Society in Turn-of-the-Century Argentina* (Stanford, CA: Stanford University Press, 2004); Norma Isabel Sánchez and Alfredo G. Kohn Loncarica, *La higiene y los higienistas en la Argentina: 1880–1943* (Buenos Aires: Sociedad Científica Argentina, 2007).

9. Anne Hardy, *The Epidemic Streets: Infectious Disease and the Rise of Preventive Medicine, 1856–1900* (Wotton-under-Edge, UK: Clarendon Press, 1993); Diego Armus, "Utopías higienicas/utopías urbanas. Buenos Aires 1920," in *Utopías urbanas: Geopolíticos del deseo en América Latina*, ed. Gisela Heffes (Frankfurt/Madrid: Verveurt, 2013), 115–130.

10. Graham Mooney, *Intrusive Interventions: Public Health, Domestic Space, and Infectious Disease Surveillance in England, 1840–1914* (Woodbridge, UK: Boydell & Brewer, 2015).

11. Adriana Alvarez, "Resignificando los conceptos de la higiene: El surgimiento de una autoridad sanitaria en el Buenos Aires de los años 80," *História, Ciências, Saúde-Manguinhos* 6, no. 2 (1999): 293–314; Susana Belmartino, *La atención médica argentina en el siglo XX: Instituciones y procesos* (Buenos Aires: Siglo Veintiuno Editores Argentina, 2005).

12. Alvarez, "Resignificando los conceptos de la higiene," 299.

13. Many scholars have emphasized that the modernization of Buenos Aires, and especially of the impoverished parts of the capital, saw fear of infection, unhygienic living conditions, and the state of morality addressed as deeply associated issues. See Diego Armus, "El descubrimiento de la enfermedad como problemo social," in *El progreso, la modernización y sus límites (1880–1916)*, ed. Mirta Zaida Lobato (Buenos Aires: Editorial Sudamericana, 2002), 515; Alvarez, "Resignificando los conceptos de la higiene"; Belmartino, *La atención médica argentina en el siglo XX*; Adriana Alvarez and Adrián Carbonetti, *Saberes y prácticas médicas en la Argentina: Un recorrido por historias de vida* (Buenos Aires: EUDEM, 2008); Norma Isabel Sánchez and Alfredo G. Kohn Loncarica, *La higiene y los higienistas en la Argentina: 1880–1943* (Buenos Aires: Sociedad Científica Argentina, 2007); Diego Armus, *The Ailing City: Health, Tuberculosis, and Culture in Buenos Aires, 1870–1950* (Durham, NC: Duke University Press, 2011); Kindon Thomas Meik, "Disease and hygiene in the construction of a nation: The Public sphere, public space, and the private domain in Buenos Aires, 1871–1910" (unpublished PhD thesis, Florida International University, 2011).

14. Luis Agote and A. J. Medina, *La peste bubonique dans la République Argentine et au Paraguay : épidémies de 1899–1900* (Buenos Aires: F. Lajouane, 1901); Myron J. Echenberg, *Plague Ports: The Global Urban Impact of Bubonic Plague, 1894–1901* (New York: New York University Press, 2007).

15. Kohl, *Higienismo Argentino.*

16. Ibid., 45.

17. José Penna, *La administración sanitaria y asistencia pública de la ciudad de Buenos Aires* (Buenos Aires: Imp. G. Kraft, 1910).

18. Kohl, *Higienismo Argentino,* 44.

19. Armus, "El descubrimiento de la enfermedad como problem social."

20. Ruggiero, *Modernity in the Flesh.*

21. Julia Rodríguez, *Civilizing Argentina: Science, Medicine, and the Modern State* (Chapel Hill: University of North Carolina Press, 2006).

22. Ibid., 41.

23. Meik, "Disease and hygiene in the construction of a nation," 25.

24. Alvarez and Carbonetti, *Saberes y prácticas médicas en la Argentina,* 62.

25. Cueto and Palmer, *Medicine and Public Health in Latin America,* 76.

26. Ruggiero, *Modernity in the Flesh,* 90.

27. Rodríguez, *Civilizing Argentina,* 42.

28. Ibid., 183.

29. Ruggiero, *Modernity in the Flesh,* 87ff.

30. Meik, "Disease and hygiene in the construction of a nation," 111.

31. Armus, *The Ailing City,* 119.

32. Armus, "Utopias higienicas," 116.

33. Echenberg, *Plague Ports,* 136.

34. Kohl, *Higienismo Argentino,* 72.

35. Penna, *La administración sanitaria,* 55.

36. Ibid.

37. Ibid., 92.

38. Ibid., 94.

39. Michael Worboys, *Spreading Germs: Disease Theories and Medical Practice in Britain, 1865–1900* (Cambridge: Cambridge University Press, 2000).

40. Bruno Latour, *The Pasteurization of France,* trans. Alan Sheridan (Cambridge, MA: Harvard University Press, 1988).

41. Penna, *La administración sanitaria*, 165.

42. Ibid., 95.

43. Ibid.

44. Armus, "El descubrimiento de la enfermedad como problema social," 533.

45. Rodríguez, *Civilizing Argentina*, 177ff.

46. Meik, "Disease and hygiene in the construction of a nation," 121.

47. Penna, *La administración sanitaria*, 72. All quotes from Penna in this chapter are the authors' translation.

48. Meik, "Disease and hygiene in the construction of a nation," 121.

49. Ibid.

50. Alvarez, "Resignificando los conceptos de la higiene."

51. Ibid.

52. Agote and Medina, *La peste bubonique*, 105.

53. Echenberg, *Plague Ports*, 137.

54. Maria Antónia Pires de Almeida, "Epidemics in the news: Health and hygiene in the press in periods of crisis," *Public Understanding of Science* 22, no. 7 (2013): 886–902.

55. Carlos Malbrán, *Apuntes sobre salud pública* (Buenos Aires, 1931).

56. Anonymous, "Otra vez la peste del Paraguay y el Departemento Nacional de Higiene," *La Semana médica* 7, no. 26 (1900): 388.

57. Agote and Medina, *La peste bubonique*, 101.

58. Carlos Malbrán, "La Peste Del Rosario," *La Semana médica* 7, no. 7 (1900): 91–94.

59. Echenberg, *Plague Ports*, 143.

60. Agote and Medina, *La peste bubonique*, 77.

61. Paul-Louis Simond, "La propagation de la peste," *Annales de l'Institut Pasteur* 62 (1898): 80–98.

62. M. Netter, "La peste durante esos ultimos anos. Sintomas—Marcha—Diagnostico," *La Semana médica* 6, no. 46 (1899): 423–428.

63. Diogenes Decoud, "La fiebre bubonica. Mecanismo del contagion," *La Semana médica* 6, no. 46 (1899): 429–30.

64. Agote and Medina, *La peste bubonique*, 211.

65. Ibid., 215.

66. Carlos Malbrán, *Apuntes sobre salud pública.*

67. Ibid., 25ff.

68. Agote and Medina, *La peste bubonique,* 99.

69. José Penna, "Consideraciones sobre la etiologia de la peste," *La Semana médica* 8, no. 28 (1901): 420.

70. Penna, *La administración sanitaria,* 150.

71. Ibid., 142.

72. Ibid.

73. Ibid.

74. Meik, "Disease and hygiene in the construction of a nation," 15. Concejo Deliberante Buenos Aires, *Memoria de la Intendencia Municipal de la Ciudad de Buenos Aires correspondiente a 1889, presentada al H. Concejo Deliberante por el Intendente Seeber* (Buenos Aires: Kraft, 1891), 129; Concejo Deliberante Buenos Aires, *Memoria presentada al Concejo Deliberante por el Intendente Municipal Señor Emilio V. Bunge; Año 1895* (Buenos Aires: Kraft, 1896), 60; Concejo Deliberante Buenos Aires, *Memoria de la Intendencia Municipal; Año 1905* (Buenos Aires: Kraft, 1906), 75.

75. Kohl, *Higienismo Argentino,* 73.

76. Penna, *La administración sanitaria,* 159.

77. Ibid., 168.

78. A. Chantemesse, "Rapport sur les expériences de destruction des rats à bord des navires au moyen de l'anhydride sulfureux liquide," in *Recueil des actes officiels et documents intéressant l'hygiène publique travaux du comité consultatif d'hygiène publique,* ed. Ministere de l'Interieur (Paris, 1905), 191–214; Gabriel Magny, *Rats et peste* (Paris : Bonvalet-Jouve, 1907).

79. Magny, *Rats et peste,* 98ff.

80. There were a number of competing machines in place at the time. A comprehensive comparison of the machines in their functionality, efficiency, and economic values is given by J. Chaine in 1932: J. Chaine, "Raport sur la destruction des rats," *Deuxième Conférence internationale et congrès colonial du rat et de la peste : Paris, 7–12 octobre 1931/documents réunis et publiés par . . . Gabriel Petit.* (Paris: Vigot, 1932), 453–463.

81. Penna, *La administración sanitaria,* 161.

82. Ibid., 168.

83. Ibid., 170.

84. Penna, *La administración sanitaria*, 173–193.

85. Penna, quoted in Kohl, *Higienismo Argentino*, 80.

86. Penna, *La administración sanitaria*, 192.

87. Ibid., 263.

88. Ibid., 265.

89. Ibid., 243.

90. Ministerio de Guerra, Inspeccion General de Sanidad, *Instrucciones sobre profilaxia humana y desinfección para el ejército Argentino* (Buenos Aires: Guillermo Kraft, 1908), 9.

91. Ibid., 49.

92. Pedro Rivero, "Saneamiento de la ciudad Buenos Aires. Deratización," *La Semana médica* 18, no. 1 (1911), 7.

93. Armus, "El descubrimiento de la enfermedad como problem social," 535.

94. Penna, *La administración sanitaria*, 119.

95. Amador, *Medicine and Nation Building in the Americas*, 8.

96. Ibid., 7ff.

97. Benchimol, *Dos micróbios aos mosquitos*; J. J. Kinyoun, "The prophylaxis of plague," *Journal of the American Medical Association* XLII, no. 4 (1904): 232–239.

98. Ministerio del Interior, Departamento Nacional de Higiene (ed.), *Tecnica teórico-práctica de la desinfección* (Buenos Aires: Guillermo Kraft, 1912).

99. Armus, "Utopias higienicas," 125.

Chapter 7

1. Anonymous, *New York Times* (October 30, 1910), 17.

2. NARA Passport Applications, January 2, 1906–March 31, 1925: 1908–1910, Roll 0095—Certificates: 14530–15430, 01 Oct 1909–27 Oct 1909, No.71, NATURALIZED application 8, 1909 14804, Thomas A. Clayton, American Embassy at Paris.

3. Dawn Day Biehler, *Pests in the City: Flies, Bedbugs, Cockroaches, and Rats* (Washington DC: Washington University Press, 2013).

4. David Lantz, "House rats and mice," *Farmers' Bulletin* 896, United States Department of Agriculture, (October 1917), 2. The bulletin was largely a reproduction

of the Farmers' Bulletin 369, but with the addition of the precautionary alarmist preface.

5. Ibid.

6. Richard H. Creel, "The rat: A sanitary menace and an economic burden," *Public Health Reports (1896–1970)* 284, no. 27 (July 4, 1913), 1403.

7. Ibid.

8. Ibid.

9. Isadore Dyer, *The Rat, the Flea, and the Plague* (1912 [offprint]. University of Alabama at Birmingham Archives), 1.

10. Creel, "The Rat," 1406.

11. Dyer, *The Rat, the Flea, and the Plague*, 4.

12. Richard H. Creel, "The rat: Its habits and their relation to antiplague measures," *Public Health Reports (1896–1970)* 28, no. 9 (February 28, 1913): 382–386.

13. "Department of Public Health, Cape Town, Cape of Good Hope, Extermination of Rats: Re, January 28, 1909." NARA RG90 Central File 1897–1923 537–544 Box 065.

14. Hannah Brown and Ann Kelly, "Material proximities and hotspots: toward an anthropology of viral hemorrhagic fevers," *Medical Anthropology Quarterly* 28, no. 2 (2014): 280–303.

15. W. J. Simpson, *A Treatise on Plague; Dealing with the Historical, Epidemiological, Clinical, Therapeutic and Preventive Aspects of the Disease* (Cambridge: Cambridge University Press, 1905), 335.

16. William David Henderson Stevenson, *Preliminary Report on the Killing of Rats and Rat Fleas by Hydrocyanic Acid Gas* (Calcutta: Superintendent Government Printing, India, 1910).

17. Ibid., 1.

18. Ibid., 8.

19. Ibid., 13.

20. C. W. Woodworth, *School of Fumigation, University of California: Held at Pomona, California* (Los Angeles: Braun Corp., 1915), 8.

21. Stevenson, *Preliminary Report on the Killing of Rats and Rat Fleas*, 8.

22. Ibid., 8.

23. Ibid., 5.

24. Stevenson mentioned a test, which had been devised by Captain Dickinson, that indicated the remaining density of hydrocyanic acid gas through colorization of a test paste: the deeper the Prussian blue, the higher the concentration of gas. Stevenson, *Preliminary Report on the Killing of Rats and Rat Fleas,* 12.

25. Ibid.

26. "US Public Health and Marine-Hospital Service, Angel Island, Cal. to The Surgeon-General, Public Health and Marine-Hospital Service, Washington D.C., Rat Quarantine report for the week ended June 15, 1912, June 20 1912"; NARA RG90 Central File, 1897–1923, 544 Box 066.

27. Ibid.

28. Ibid.

29. R. H. von Ezdorf, "The occurrence of plague in Habana and the measures adopted for its control and eradication," *Public Health Reports (1896–1970)* 27, no. 42 (October 18, 1912): 1697–1702.

30. Ibid.

31. "US Public Health and Marine-Hospital Service, Office of Medical Officer in Command, New Orleans Quarantine Station, 'Annual Report of "Rat Quarantine,"' Port of New Orleans, for the year ending June 30th 1912"; NARA RG90 Central File, 1897–1923, 544 Box 066.

32. "Health Officer, Port of New York to John A. Dix, Albany N.Y. August 3 1912"; NARA RG90 Central File, 1897–1923, 544 Box 066.

33. Ibid.

34. Ibid.

35. "US Public Health and Marine-Hospital Service, Office of Medical Officer in Command, New Orleans Quarantine Station to Mr E. Ferrer, The New York & Porto Rico Steamship Company, San Juan, Porto Rico, August 3 1912"; NARA RG90 Central File, 1897–1923, 544 Box 066.

36. Ibid.

37. "US Public Health and Marine-Hospital Service, Angel Island, Cal., Rat Quarantine to The Surgeon-General, Public Health and Marine-Hospital Service, Washington D.C., July 30 1912"; NARA RG90 Central File, 1897–1923, 544 Box 066.

38. See, for example, "International Mercantile Marine Company Steamship Department J. H. Thomas to Bureau of P.H and M.hS, July 27 1909 Rel. fumigation of ships for the destruction of rats"; NARA RG90 Central File 1897–1923 537–544 Box 065.

39. "Munson Steamship Line, F. C. Munson to to Bureau of P.H and M.hS, Relative to the fumigation of ships for the destruction of rats, July 23 1909"; NARA RG90 Central File 1897–1923 537–544 Box 065.

40. Anonymous, "Ship rats and plague," *Public Health Reports (1896–1970)* 29, no. 16 (April 17, 1914): 927–928.

41. "Red 'D"' Line Steamers, Boulto, Bliss & Dallett to Bureau of P.H and M.hS, Relative to the fumigation of ships for the destruction of rats July 22 1909"; NARA RG90 Central File 1897–1923 537–544 Box 065.

42. Richard H. Creel, "Outbreak and suppression of plague in Porto Rico: An account of the course of the epidemic and the measures employed for its suppression by the United States Public Health Service," *Public Health Reports (1896–1970)* 28, no. 22 (May 30, 1913): 1050–1070.

43. S. B. Grubbs, "Fumigation of vessels from plague-infected ports. Observations with especial reference to the necessity for fumigating crates and similar cargo," *Public Health Reports (1896–1970)* 38, no. 2 (January 12, 1923): 59–63.

44. Richard Campanella, "The battle against bubonic plague: 100 years ago, New Orleans waged war on rats," *The Times Picayune* (May 8, 2014). http://www.nola.com/homegarden/index.ssf/2014/08/one_hundred_years_ago_as_war_b.html.

45. Norman Roberts, "Cyanide fumigation of ships: Method used at New Orleans," *Public Health Reports (1896–1970)* 29, no. 50 (December 11, 1914), 3321.

46. Ibid., 3322.

47. Ibid., 3324.

48. Richard H. Creel, F. M. Faget, and W. D. Wrightson, "Hydrocyanic acid gas: Its practical use as a routine fumigant," *Public Health Reports (1896–1970)* 30, no. 49 (December 3, 1915): 3537–3550.

49. Ibid., 3538.

50. Ibid., 3539.

51. Ibid., 3546.

52. Ibid., 3547.

53. Further technical innovations, including the artificial ventilation (for example, with electric fans), after HCN fumigation would be developed in 1917; see: S. B. Grubbs, "Ventilation after fumigation: Artificial ventilation of ships after fumigation with hydrocyanic acid gas," *Public Health Reports (1896–1970)* 32, no. 42 (October 19, 1917): 1757–1761.

54. Creel et al., "Hydrocyanic acid gas," 3547.

55. Richard H. Creel and Friench Simpson, "Rodent destruction on ships: A report on the relative efficiency of fumigants as determined by subsequent intensive trapping over a period of one year," *Public Health Reports (1896–1970)* 32, no. 36 (September 7, 1917), 1446.

56. Ibid.

57. Edmund Russell, *War and Nature: Fighting Humans and Insects with Chemicals from World War I to* Silent Spring (Cambridge: Cambridge University Press, 2001).

58. L. F. Haber, *The Poisonous Cloud: Chemical Warfare in the First World War* (Oxford: Oxford University Press, 1986), 98.

59. Ibid.

60. R. C. Roark, *A Bibliography of Chloropicrin 1848–1932* (Washington, DC: US Department of Agriculture, Miscellaneous Publication No. 176, 1934), 2.

61. Abdul-Magid Fayed, *Le dératisation antipesteuse* (Toulouse, France: Impr. des Artes, 1928).

62. G. Bertrand, "Sur la haute toxicité de la chloropicrine vis-à-vis de certains animaux inferieurs et sur la possibilité d'emploi de cette substance comme parasiticide," *Compte rendu Academie des Sciences* 168 (1919), 742.

63. Paul Secques, *La dératisation par la chloropicrine* (Toulouse, France: Impr. P. Julia, 1937).

64. Roark, *A Bibliography of Chloropicrin 1848–1932*, 2.

65. Indicatively, in Marseilles, the Hygiene Council approved the following: S.I.C. du Midi process, February 1923; The Allienne apparatus, January 1926; Santos apparatus, July 1929; Worms apparatus, July 1928; Edde Process, March 1931; François Rech, *La dératisation à Marseille* (Marseille: Société anonyme du sémaphore de Marseille, 1932.)

66. "C. M. Fauntelroy to the Surgeon General US Public Health Service, Sept. 9 1922"; NARA RG90 Central File, 1897–1923, 544 Box 066.

67. Grubbs, "Fumigation of vessels from plague-infected ports." Lighters themselves were deemed to be of great importance as means of transport of rats from ship to the harbor, and thus needing fumigation; on the implementation of this in Hong Kong, see, for example, "C. M. Fauntelroy to the Surgeon General US Public Health Service, October 12 1922"; NARA RG90 Central File, 1897–1923, 544 Box 066.

68. Grubbs, "Fumigation of vessels from plague-infected ports," 61.

69. Ibid.

70. Ibid.

71. Grubbs to Surgeon General, February 21 1922 Rat Return & Examination of Same at this Station; NARA RG90.

72. Ibid.

73. Grubbs was reprimanded for making this optional.

74. See Clifford Rush Eskey and V. H. Haas, "Plague in the western part of the United States: infection in rodents, experimental transmission by fleas, and inoculation tests for infection," *Public Health Reports (1896–1970)* 54, no. 32 (August 11, 1939): 1467–1481.

75. John D. Long, "A squirrel destructor: An efficient and economical method of destroying ground squirrels," *Public Health Reports (1896–1970)* 27, no. 39 (1912): 1594–1596.

76. W. Allen Daley, "Deratisation and the International Sanitary Convention of 1926," *Perspectives in Public Health* 49, no. 6 (1928), 355.

77. Colombani and Herminier, "Sur la fumigation des navires à l'aide d'anhydride sulfureux-sulfurique ou d'acide cyanhidrique," in *Première conférence internationale du rat, Paris—Le Havre 16–22 Mai 1928,* ed. Gabriel Petit (Paris: Vigot frères, 1928), 186–189.

78. Discussion, M. le Dr. A. Lutratio, in Ibid.

79. The delegates enjoyed a demonstration of this machine and "Gaz Sanos" on the ship *Texas* of the Compagnie Générale Transatlantique showcasing the extraordinary precision of HCN. René Gissy, "La conférence international du rat au Havre," in *Première conférence internationale du rat, Paris—Le Havre 16–22 Mai 1928,* ed. Gabriel Petit (Paris: Vigot frères, 1928), 313–324.

80. Gaston Doumergue, "Le décret du 8 aout 1929, relatif à la dératisation des navires (journal officiel du 14 aout 1929)," in *Première conférence internationale du rat, Paris—Le Havre 16–22 Mai 1928,* ed. Gabriel Petit (Paris: Vigot frères, 1928), 352–353.

81. R. Chamaillard, "L'emploi de l'acide cyanhydrique pour la dératisation," in *Deuxième conférence internationale et congrès colonial du rat et de la peste, Paris 7–12 Octobre 1931,* ed. Gabriel Petit (Paris: Vigot frères, 1931), 500–511.

82. To give the example of Marseilles, there the following were approved: Sanos process, May 1926; Autran and Galardi apparatus, March 1927; Zyklon process, March 1929; Galardi process, March 1930; Sansone process, March 1930; Vicent process, March 1930; Edde-Hygroma process, December 1930; Rech, *La dératisation à Marseille.*

83. L. F. Hirst, *The Protection of the Interior of Ceylon from Plague with Special Reference to the Fumigation of Plague-Suspect Imports* (Colombo, Sri Lanka: Municipal Printing Office, 1931).

84. Ibid., 29.

85. W. Glen Liston and S. N. Goré, "The fumigation of ships with Liston's cyanide fumigator," *Journal of Hygiene* 21, no. 3 (May 1923), 200.

86. Ibid. Other HCN-based machines like the "Hyproma" followed Liston's invention and were discussed in the Second International Rat Conference; Chamaillard, "L'emploi de l'acide cyanhydrique pour la dératisation."

87. Hirst, "The Protection of the Interior of Ceylon from Plague," 30.

88. Ibid.

89. Paul Weindling, "The uses and abuses of biological technologies: Zyklon B and gas disinfestation between the First World War and the Holocaust," *History and Technology* 11, no. 2 (1994): 291–298.

90. Ibid., 294.

91. Ibid., 295.

92. C. V. Akin and G. C. Sherrard, "Fumigation with cyanogen products. Report of experiments conducted with cyanogen products used in the fumigation of vessels for quarantine purposes at the New York Quarantine Station, Rosebank, Staten Island, N.Y.," *Public Health Weekly Reports* (1896–1970) 43, no. 41 (October, 12 1928): 2643–2670.

93. Ibid., 2669.

94. Ibid., 2670.

95. Chamaillard, "L'emploi de l'acide cyanhydrique."

96. C. L. Williams, "The air jet hydrocyanic acid sprayer," *Public Health Reports (1896–1970)* 46, no. 30 (July 24, 1931), 1755.

97. On extensive studies of rat harborages and how rats escape fumigation, conducted in New York Quarantine Station. see C. L. Williams, "Notes on the fumigation of vessels. Preliminary inspection, how rats escape, increased periods of exposure, and other miscellaneous notes," *Public Health Reports (1896–1970)* 46, no. 50 (December 11, 1931): 2973–2980.

98. Williams, "The air jet hydrocyanic acid sprayer." For a discussion of a "Zyklon pump," see C. L. Williams, "The fumigation of loaded ships," *Public Health Reports (1896–1970)* 46, no. 31 (July 31, 1931): 1823–1836.

99. J. R. Ridlon, "Some aspects of ship fumigation," *Public Health Reports (1896–1970)* 46, no. 27 (July 3, 1931), 1572.

100. Williams, "The air jet hydrocyanic acid sprayer," 1751.

Bibliography

Archival Sources

Archives nationales d'outre-mer (ANOM)
  INDO GGI 4416

The British Library
  IOR/R/20/A/2596 File 118/7 Plague: Rat destruction.

The Countway Library, Harvard University
  Leatherbee, Mrs. Albert T., and the Women's Municipal League of Boston. *Plague Conditions in Boston*, 1921. [Pamphlet]

Hong Kong Government Gazette
  GA 1895 no.146; Medical Report on the Epidemic of Bubonic Plague in 1894 (incorporating J. A. Lowson, "The Epidemic of Bubonic Plague in Hong Kong, 1894," April 13, 1895): 369–422.

The Matas Medical Library, Tulane Universiy, New Orleans
  Louisiana Board of Health. *Annual Reports of the Louisiana State Board of Health*. New Orleans, 1875–1900.

National Archives (UK)
  MH 19/274, Plague Destruction of Rats on Ships

  Colonial Office. Original Correspondence: Hong Kong 1841–1951; Series 129/263, CO17303.

The National Archives and Records Administration (USA)
  NARA RG90 Central File 1897–1923 537–544.

  Passport Applications, January 2, 1906—March 31, 1925

Parliamentary Papers (UK)
  Reports from the Committees of the House of Commons, Reprinted by Order of the House, Vol. XIV, Miscellaneous Reports, 1793–1802.

Wellcome Library
Wellcome Library, Reference number b19884217, Port of London Sanitary Committee, Annual Report of the Medical Officer to 31st December 1902.

University of Alabama at Birmingham Archives
Dyer, Isadore. *The Rat, the Flea, and the Plague*. 1912. [Offprint]

US Patents Registry
Clayton, Thomas A. Method and Apparatus for Fumigating and Extinguishing Fires in Closed Compartments. *Patent US633807*. Filed June 7, 1899, granted September 26, 1899.

Olliphant, S. R., and T. A. Clayton. Fumigation Apparatus, *Patent US490981*, filed January 25, 1892, granted January 31, 1893.

Primary Published Sources

Anonymous. No title. *The Times Democrat* (New Orleans, LA), April 11, 1890, 11.

Anonymous. No title. *St. Landry Democrat*, April 19, 1890, 1.

Anonymous. "Board of Health. An address to the public as to its policy and plans." *The Daily Picayune* (New Orleans, LA), April 23, 1890, 6.

Anonymous. "Trip to quarantine. Official visit of inspection by the Board of Health and legislative committees." *The Daily Picayune* (New Orleans, LA), June 2, 1890, 3.

Anonymous. No title. *The Times-Democrat* (New Orleans, LA), May 17, 1891, 3.

Anonymous. No title. *St. Landry Democrat*, June 6, 1891, 4.

Anonymous. No title. *The Times-Democrat* (New Orleans, LA) July 22, 1891, 3.

Anonymous, *Biographical and Historical Memoirs of Louisiana*. Chicago: The Goodspeed Publishing Company, 1892.

Anonymous. No title. *The Daily Picayune* (New Orleans, LA), July 24, 1894, 11.

Anonymous. City Directory, New Orleans, Louisiana, 1895.

Anonymous. "A case in point." *Washington Post* (1877–1922), December 17, 1897, 6.

Anonymous. "Bulletin. La peste bubonique." *Revue médico-pharmaceutique* 11, no. 8 (August 15, 1898): 101–102.

Anonymous. "La peste." *Gazette médicale d'Orient* 44, nos. 19 and 20) (November 30 and December 15, 1899): 266–267.

Anonymous. "Otra vez la peste del Paraguay y el Departemento Nacional de Higiene." *La Semana médica* 7, no. 26 (1900): 388–389.

Anonymous. "Memoire sur l'influence des rats dans la propagation de la peste." *Revue médico-pharmaceutique* 13, no. 22 (November 15, 1900): 253–257.

Anonymous. "Destruction des rats à bord des navires." *Revue médico-pharmaceutique* 13, no. 23 (December 1, 1900): 266.

Anonymous. "The Plague. The destruction of rats on shipboard." *The British Medical Journal* 1, no. 2106 (May 11, 1901): 1170.

Anonymous. No title. *La Revue hebdomadaire* 4, no. 47 (October 19, 1901): 2–4.

Anonymous. "Le lazaret du Frioul." *L'actualité* 91 (October 20, 1901): 1.

Anonymous. "Discussions: Sur la police sanitaire maritime et le sejour du *Sénégal* au lazaret du Frioul (Séance 5 novembre 1901)." *Bulletin de l'Académie nationale de médecine* 3, no. XLVI (1901): 488–503.

Anonymous. "La peste à bord du Sénégal." *Petit Journal* (December 1, 1901): 264.

Anonymous. [advertisement] *Industrial Management* 21 (1901): 998–999.

Anonymous. "The destruction of rats on ship-board." *The British Medical Journal* 1, no. 2144 (February 1, 1902): 294–295.

Anonymous. *The St. Paul Globe* (Minnesota), February 7, 1904, 21.

Anonymous. "Décret concernant la destruction des rats à bord des navires provenant de pays contaminés de peste (4 mai 1906)." *Bulletin officiel du ministère de l'intérieur* 5 (1906): 261–263.

Anonymous. "Method of fumigation of vessels at Hamburg." *United States Naval Medical Bulletin* 3 (1909): 368–370.

Anonymous. No title. *New York Times*, October 30, 1910, 17.

Anonymous. "Ship rats and plague." *Public Health Reports (1896–1970)* 29, no. 16 (April 17, 1914): 927–928.

Anonymous. "Elimination of the rat." *Boston Medical and Surgical Journal* 174, no. 2 (October 19, 1916): 576.

Actes officiels. "Destruction des rats à bord des navires provenant de pays contaminés de peste, avant déchargement, décret du 21 septembre 1903." *Recueil des actes officiels et documents intéressant l'hygiène publique (Travaux du Conseil supérieur d'hygiène publique de France)* 33 (1903): 106–107.

Agote, Luis, and A. J. Medina. *La peste bubonique dans la République Argentine et au Paraguay : Épidémies de 1899–1900.* Buenos Aires: F. Lajouane, 1901.

American Public Health Association, Committee on Disinfectants and Royal College of Physicians of Edinburgh. *Report of the Committee on Disinfectants of the American Public Health Association.* Baltimore: Printed for the Committee, 1885.

Akin, C. V., and G. C. Sherrard. "Fumigation with cyanogen products. Report of experiments conducted with cyanogen products used in the fumigation of vessels for quarantine purposes at the New York Quarantine Station, Rosebank, Staten Island, N.Y." *Public Health Weekly Reports* (1896–1970) 43, no. 41 (October 12, 1928): 2643–2670.

Apéry, Pierre. "Lettre ouverte, Reponse à M. Desguin, L'anhydride carbonique peut-il debarasser des rats les cales des bateaux?" *Revue médico-pharmaceutique* 13, no. 8 (April 15, 1900): 86–89.

Apéry, Pierre. "Bulletin épidémiologique." *Revue médico-pharmaceutique* 14, no. 16 (August 15, 1901): 181.

Apéry, Pierre. "Moyen de destruction des rats à bord des bateaux surtout en temps d'épidémie de peste." *Revue médico-pharmaceutique* 12, no. 10 (October 15, 1899): 137–138.

Apéry, Pierre. "Bulletin épidémiologique. L'utilité des quarantaines." *Revue médico-pharmaceutique* 14, no. 21 (November 1, 1901): 241.

Apéry, Pierre. "Les quarantaines supprimées." *Revue médico-pharmaceutique* 14, no. 23 (December 1, 1901): 265.

Apéry, Pierre. "Note sur l'emploi de l'anhydride carbonique pour la destruction des rats dans les cales des navires." *Revue médico-pharmaceutique* 15, no. 1 (January 1, 1902): 1–3.

Apéry, Pierre. "De l'inefficacité des systems à base d'anhydride sulfureux employés jusqu'à ce jour pour la destruction des rats dans des navires." *Revue médico-pharmaceutique* 15, no. 4 (February 15, 1902): 36–39.

Balilis, S. *Contribution à l'étude des mesures sanitaires contre la propagation de la peste.* Constantinople: Impr. A. Zellich fils, 1901.

Bacot, A. W., and C. J. Martin. "Observations on the mechanism of the transmission of plague by fleas." *Journal of Hygiene*, Plague Suppl. 3 (1914): 423–439.

Barthélemy, and Dr. Georges Varenne. *Manuel d'hygiène navale: À l'usage des capitaines, des officiers et des élèves de la marine marchande.* Paris: Challamel, 1907.

Bartholow, Roberts. *The Principles and Practice of Disinfection.* Cincinnati: R. W. Carroll & Co., 1867.

Bertot, Jean. *Au lazaret. Souvenirs de quarantaine.* Tours, France: Deslis frères, 1902.

Bertrand, G. "Sur la haute toxicité de la chloropicrine vis-à-vis de certains animaux inferieurs et sur la possibilité d'emploi de cette substance comme parasiticide." *Compte rendu Academie des Sciences* 168 (1919): 742–744.

Bonkowski, C. "De l'emploi de l'anhyhride carbonique pour la destruction des rats dans les cales des navires." *Revue médico-pharmaceutique* 14, no. 23 (December 1 1901): 264–267.

Borel, Frédéric. *Étude d'hygiène internationale. Choléra et peste dans le pèlerinage musulman, 1860–1903*. Paris: Masson, 1904.

Bucqcoy, Jules. "Une quarantaine au Frioul (séance du 29 octobre 1901)." *Bulletin de l'Académie nationale de médecine* 3, no. XLVI (1901): 422–436.

Cabanelles, Miguel. *Defensa de las fumigaciones ácido-minerales*. Madrid: Repullés, 1814.

Carmichael Smyth, James. *Description of the Jail Distemper as it Appeared amongst the Spanish Prisoners at Winchester, in the Year 1780*. London: J. Johnson, 1795.

Chaine, J. "Raport sur la destruction des rats." In *Deuxième conférence internationale et congrès colonial du rat et de la peste: Paris, 7–12 octobre 1931*, ed. Gabriel Petit, 453–463. Paris: Vigot frères, 1932.

Chamaillard, R. "L'emploi de l'acide cyanhydrique pour la dératisation." In *Deuxième conférence internationale et congrès colonial du rat et de la peste, Paris 7–12 Octobre 1931*, ed. Gabriel Petit, 500–511. Paris: Vigot frères, 1931.

Chantemesse, A. "Rapport sur les expériences de destruction des rats à bord des navires au moyen de l'anhydride sulfureux liquide." *Recueil des actes officiels et documents intéressant l'hygiène publique travaux du Comite Consultatif d'Hygiene Publique* (1905): 191–214.

Colombani and Herminier. "Sur la fumigation des navires à l'aide d'anhydride sulfureux-sulfurique ou d'acide cyanhidrique." *Première conférence internationale du rat, Paris—Le Havre 16–22 Mai 1928*, ed. Gabriel Petit, 186–189. Paris: Vigot frères, 1928.

Concejo Deliberante Buenos Aires. *Memoria de la Intendencia Municipal de la Ciudad de Buenos Aires correspondiente a 1889, presentada al H. Concejo Deliberante por el Intendente Seeber*. Buenos Aires: Kraft, 1891.

Concejo Deliberante Buenos Aires, *Memoria presentada al Concejo Deliberante por el Intendente Municipal Señor Emilio V. Bunge; Año 1895*. Buenos Aires: Kraft, 1896.

Concejo Deliberante Buenos Aires, *Memoria de la Intendencia Municipal; Año 1905*. Buenos Aires: Kraft, 1906.

Conseil Quarantenaire d' Égypte. "Rapport adressé à la presidence du Conseil Quarantinaire d' Égypte par M. le directeur P. I. de l'Office Quarantinaire d'Alexandrie sur la destruction des rats à bord des navires." *Recueil des Travaux du Comité Consultatif d'Hygiène Publique de France et des Actes Officiels de l'Administration Sanitaire* 31 (1901): 513–516.

Conseil général d'administration des hospices. *Descriptions des appareils à fumiga-tions, établis, sur les dessins de M. D'Arcet, à l'hôpital Saint-Louis, en 1814 et successive-ment dans plusieurs Hôpitaux de Paris, pour le traitement des Maladies de la peau*. Paris: Impr. des Hospices civils, 1818.

Corbin, Alain. *Le miasma et la jonquille. L'odorat et l'imaginaire social aux XVIIIe et XIXe siècles*. Paris: Champs histoire, 2016 [1982].

Cozzonis Effendi. *Rapport sur la manifestation pestilentielle à Djeddah en 1898, suivi d'une esquisse sur les conditions generales de la dite ville*. Istanbul: Impr. Osmanié, 1898.

Creel, Richard H. "The rat: Its habits and their relation to antiplague measures." *Public Health Reports (1896–1970)* 28, no. 9 (February 28 1913): 382–386.

Creel, Richard H. "The rat: A sanitary menace and an economic burden." *Public Health Reports (1896–1970)* 284, no. 27 (July 4, 1913): 1403–1408.

Creel, Richard H., F. M. Faget, and W. D. Wrightson. "Hydrocyanic acid gas: Its practical use as a routine fumigant." *Public Health Reports (1896–1970)* 30, no. 49 (December 3, 1915): 3537–3550.

Creel, Richard H. and Friench Simpson. "Rodent destruction on ships: A report on the relative efficiency of fumigants as determined by subsequent intensive trapping over a period of one year." *Public Health Reports (1896–1970)* 32, no. 36 (September 7, 1917): 1445–1450.

Creel, Richard H. "Outbreak and suppression of plague in Porto Rico: An account of the course of the epidemic and the measures employed for its suppression by the United States Public Health Service." *Public Health Reports (1896–1970)* 28, no. 22 (May 30, 1913): 1050–1070.

David, M. E., and G. Duriau. "État actuel de la désinfection des navires. Carbonica-tion, sulfuration (procédé Clayton)." *Revue d'hygiène et de police sanitaire* 25 (1903): 500–521.

Daley, W. Allen. "Deratisation and the International Sanitary Convention of 1926." *Perspectives in Public Health* 49, no. 6 (1928): 354–363.

de Cerro, Jean. *Observations pratiques sur les fumigations sulfureuses*. Vienna: Charles Gerold 1819.

Decoud, Diogenes. "La fiebre bubonica. Mecanismo del contagio." *La Semana médica* 6, no. 46 (1899): 429–430.

de Gimbernat y Grassot, Carlos. *Instruction sur les moyens propres à prévenir la conta-gion des fièvres épidémiques*. Strasbourg, France: F. G. Levrault, 1814.

Delacour. "La peste." *Revue médico-pharmaceutique* 11, no. 8 (August 15, 1898): 97–99.

Dougall, John "Disinfection by acids." *British Medical Journal* 2, no. 984 (November 8, 1879): 726–728.

Doumergue, Gaston. "Le décret du 8 aout 1929, relatif à la dératisation des navires (journal officiel du 14 aout 1929)." In *Première conférence internationale du rat, Paris—Le Havre 16–22 Mai 1928,* ed. Gabriel Petit, 352–353. Paris: Vigot frères, 1928.

Dubos, René. *Louis Pasteur, Free Lance of Science.* Boston: Little, Brown and Company, 1950.

Duriau, Gustave. "Quelques mots sur l'appareil Clayton." *Mémoires de la Société Dunkerquoise pour l'encouragement des sciences, des lettres et des arts* 36 (1902): 401–406.

Emerson, G. "Cases illustrative of the efficacy of sulphurous fumigations, in certain cases—with preliminary remarks." *Philadelphia Journal of the Medical and Physical Sciences* 3 (1821): 126–145.

Eskey, Clifford Rush, and V. H. Haas. "Plague in the western part of the United States: infection in rodents, experimental transmission by fleas, and inoculation tests for infection." *Public Health Reports (1896–1970)* 54, no. 32 (August 11, 1939): 1467–1481.

Fayed, Abdul-Magid. *Le dératisation antipesteuse.* Toulouse, France: Impr. des Artes, 1928.

Fenner, Erasmus Darwin. *History of the Epidemic Yellow Fever at New Orleans, La., in 1853.* New York: Hall, Clayton, 1854.

Freeport Sulphur Company. *4,000 Years of Yellow Magic; The Story of a Basic Element, Widely Used, Little Understood, Which for 40 Centuries Has Been Shaping the History of the World.* Freeport, TX; Port Sulphur, LA: [n.p.], 1900.

Gagnière, Sylvain. *La désinfection des caveaux d'églises après les grandes épidémies de peste.* Avignon, France: Impr. Rulliere frères, 1943.

Galès, J. C. *Mémoires et rapports sur les fumigations sulfureuses appliquées au traitement des affections cutanées et de plusieurs autres maladies.* Paris: L'Impremerie Royale, 1816.

Gissy, René. "La conférence international du rat au Havre." In *Première conférence internationale du rat, Paris—Le Havre 16–22 Mai 1928,* ed. Gabriel Petit, 313–324. Paris: Vigot frères, 1928.

Grubbs, S. B. "Ventilation after fumigation: Artificial ventilation of ships after fumigation with hydrocyanic acid gas." *Public Health Reports (1896–1970)* 32, no. 42 (October 19, 1917): 1757–1761.

Grubbs, S. B. "Fumigation of vessels from plague-infected ports. Observations with especial reference to the necessity for fumigating crates and similar cargo." *Public Health Reports (1896–1970)* 38, no. 2 (January 12, 1923): 59–63.

Guyton de Morveau, Louis-Bernard. *Traité des moyens de désinfecter l'air*. Paris: Bernard, 1801.

Guyton de Morveau, Louis-Bernard. *A Treatise on the Means of Purifying Infected Air*, translated by R. Hall. London: J. & E. Hodson, 1802.

Henderson Stevenson, William David. *Preliminary Report on the Killing of Rats and Rat Fleas by Hydrocyanic Acid Gas*. Calcutta: Superintendent Government Printing, India, 1910.

Henry, William. "Experiments on the disinfecting powers of increased temperatures, with a view to the suggestion of a substitute for quarantine." *Philosophical Magazine* 2nd ser. 10, no. 9 (1832): 363–369.

Henry, William. "Further experiments on the disinfecting powers of increased temperatures." *Philosophical Magazine* 2nd ser. 11, no. 61 (1832): 22.

Henry, William. "Letter from Dr. Henry on a modified disinfecting apparatus." *Philosophical Magazine* 2nd ser. 11, no. 63 (1832): 205.

Hirst, L. F. *The Protection of the Interior of Ceylon from Plague with Special Reference to the Fumigation of Plague-Suspect Imports*. Colombo, Sri Lanka: Municipal Printing Office, 1931.

Holt, Joseph. *Quarantine and Commerce, Their Antagonism Destructive to the Prosperity of City and State*. New Orleans: Graham and Sons, 1884.

Holt, Joseph. "The sanitary protection of New Orleans, municipal and maritime." *The Sanitarian*, no. 194 (1886): 37–49.

Holt, Joseph. "Quarantine operations." In *Annual Report of the Louisiana Board of Health, 1885 and 1886*, ed. Louisiana Board of Health, 23–38. New Orleans: Louisiana Board of Health, 1887.

Holt, Joseph. *An Epitomized Review of the Principles and Practice of Maritime Sanitation*. New Orleans: Graham, 1892.

Howard, John. *The State of the Prisons in England and Wales: With Preliminary Observations and an Account of Some Foreign Prisons*. Cambridge: Cambridge University Press, 2013 [1777].

Kelaiditis, C. "Opinion [sur] S. Balis, 'Contribution à l'étude des mesures sanitaires contre la propagation de la peste.'" *Revue medico-pharmaceutique* 15, no. 1 (January 1, 1902): 18–19.

Khayat, Rechit. *Prophylaxie de la peste par la destruction des insectes et des rongeurs*. Paris: Jules Rousset, 1902.

Koch, Robert. "Ueber Desinfektion." *Mittheilungen aus dem Kaiserlichen Gesundheitsamte* 1 (1881): 287–338.

Koch, Robert. [Interview]. *Hamburger Freie Presse*, November 26, 1892.

Köslin (Regierungsbezirk). *Amts-Blatt der preußischen Regierung zu Köslin.* Köslin, Prussia: C. G. Hendess, 1831.

Lalouette, Pierre. *A New Method of Curing the Venereal Disease by Fumigation: Together with Critical Observations on the Different Methods of Cure and an Account of Some New and Useful Preparations of Mercury.* London: J. Wilkie, 1777.

Lalouette, Pierre. *Nouvelle méthode de traiter les maladies vénériennes, par la fumigation: avec les procès-verbaux des Guérisons opérées par ce moyen.* Paris: Merigot, 1776.

Langlois J.-P., and A. Loir. "La destruction des rats à bord des bateaux comme mesure prophylactique contre la peste." *Revue d'hygiène et de police sanitaire* 24 (1902): 411–422.

Lantz, David. "House rats and mice." *Farmers' Bulletin* 896, United States Department of Agriculture, October 1917.

Lanz, David E. *How to Destroy Rats.* Washington: Government Printing Office, 1909.

Lind, James. *An Essay on the Most Effective Means of Preserving the Health of Seamen in the Royal Navy*, 2nd edition. London: D. Wilson, 1762.

Lind, James. *Two Papers on Fevers and Infections which Were Read before the Philosophical and Medical Society in Edinburgh.* London: D. Wilson, 1763.

Lister, Joseph. "On the antiseptic principle in the practice of surgery." *The Lancet* 2, no. 2299 (September 21, 1867): 353–356

Liston, W. Glen, and S. N. Goré. "The fumigation of ships with Liston's cyanide fumigator." *Journal of Hygiene* 21, no. 3 (May 1923): 199–219.

Long, John D. "A squirrel destructor: An efficient and economical method of destroying ground squirrels." *Public Health Reports (1896–1970)* 27, no. 39 (1912): 1594–1596.

Louisiana Board of Health. *The System of Maritime Sanitation Inaugurated and Brought to Its Present State of Perfection by the Louisiana State Board of Health.* Chicago, 1893.

Louisiana Board of Health. *The Louisiana State Board of Health, Its History and Work, with a Brief Review of Health Legislation and Maritime Quarantine in Louisiana.* St. Louis, MO, 1904.

Luthy, David. *Notice sur les fumigations sulfureuses, appliquées au traitement des affections cutanées, et de plusieurs autres maladies, avec la description exacte d'un appareil pour les administrer.* Fribourg, Switzerland: Aloyse Eggendorffer, 1818.

Magny, Gabriel. *Rats et peste.* Paris: Paris: Bonvalet-Jouve, 1907.

Malbrán, Carlos. "La peste del Rosario." *La Semana médica* 7, no. 7 (1900): 91–94.

Malbrán, Carlos. *Apuntes sobre salud pública*. Buenos Aires, 1931.

Mangin, Arthur. *L'air et le monde aérien*. Tours, France: Alfred Mame et fils, 1865.

Manson, Patrick. "Dr. Manson's report on the health of Amoy for the half-year ended 31st March 1878." *Customs Gazette, Medical Reports* 2 (January–March, 1878): 25–27.

Mayhew, Henry. *London Labour and the London Poor*, 4 vols. London: Dover, 1860–1861.

Mazaraky, G. *Le rôle des rats dans la propagation de la peste*. Thèse de méd., Fac. de Médecine de Paris. Paris: Vigot Frères, 1901.

Ministère des Affaires Étrangère. *Conférence sanitaire internationale de Paris: 10 octobre-3 décembre 1903, procès-verbaux*. Paris: Impr. Nationale, 1904.

Ministerio de Guerra, Inspeccion General de Sanidad. *Instrucciones sobre profilaxia humana y desinfección para el ejército Argentino*. Buenos Aires: Kraft, 1908.

Ministerio del Interior, Departamento Nacional de Higiene (ed.), *Tecnica teórico-práctica de la desinfección*. Buenos Aires: Guillermo Kraft, 1912.

Monod, Henri. "Incident du paquebot 'Sénégal' retenu au lazaret de Frioul le 20 september 1901 comme infecté de peste." *Recueil des travaux du Comité Consultatif d'Hygiène Publique de France et des Actes Officiels de l'Administration Sanitaire* 31 (1901): 181–201.

Mordtmann, Andreas David. "Bulletin sanitaire." *Gazette médicale d'Orient* 45, no. 7 (May 30, 1900): 101–102.

Mordtmann, Andreas David. "Bulletin épidémiologique." *Gazette médicale d'Orient* 45, no. 14 (September 15, 1900): 267–268.

Mordtmann, Andreas David. "Bulletin sanitaire." *Gazette médicale d'Orient* 45, no. 23 (January 31, 1901): 459–460.

Mordtmann, Andreas David. "Bulletin épidémiologique." *Gazette médicale d'Orient* 46, no. 18 (November 15, 1901): 817–819.

Netter, M. "La Peste durante esos ultimos anos. Sintomas—Marcha—Diagnostico." *La Semana médica* 6, no. 46 (1899): 423–28.

Nocht, Bernhard, Emil von Behring, Joseph Brix, E. Pfuhl, and the Royal College of Physicians of Edinburgh. *Die Bekämpfung der Infectionskrankheiten: Hygienischer Theil*. Leipzig, Germany: G. Thieme, 1894.

Nocht, Bernhard. "Ueber die Abwehr der Pest." *Archiv fuer Schiffs und Tropen Hygiene* 1, no. 2 (1897): 91–101.

Nocht, Bernhard, and G. Giemsa. "Über die Vernichtung von Ratten an Bord von Schiffen. Als Massregel gegen die Einschleppung der Pest." *Arbeiten aus dem Kaiserlichen Gesundheitsamte* XX, no. 1 (1904): 91–110.

Pellisier, Joseph. *La peste au Frioul, lazaret de Marseille en 1900 et 1901*. Paris: Steinheil, 1902.

Penna, José. "Consideraciones sobre la etiologia de la peste." *La Semana médica* 8, no. 28 (1901): 407–420.

Penna, José. *La administración sanitaria y asistencia pública de la ciudad de Buenos Aires*. Buenos Aires: Imp. G. Kraft, 1910.

Père Maurice de Toulon. *Le Capucin charitable enseignant le méthode pour remédier aux grandes misères que la peste a coutume de causer parmi les peoples*. Lyon: Bruyet: 1722.

Perry, A. W. "Effectual external regulations without delay to commerce." *Public Health Papers and Reports* 1 (1873): 437–440.

Proust, Adrien. *La défense de l'Europe contre la peste et la conférence de Venise de 1897*. Paris: Masson, 1897.

Proust, A., and Paul Faivre. "Rapport général sur les maladies pestilentiels exotiques (peste, fièvre jaune, cholera) en 1901." *Recueil des Travaux du Comité Consultatif d'Hygiène Publique de France et des Actes Officiels de l'Administration Sanitaire* 31 (1901): 203–354.

Proust, A., and Paul Faivre. "Rapport sur different procédés de destruction des rats à bord des navires." *Recueil des Travaux du Comité Consultatif d'Hygiène Publique de France et des Actes Officiels de l'Administration Sanitaire* 33 (1903): 335–381.

Rech, François. *La dératisation à Marseille*. Marseille: Société anonyme du sémaphore de Marseille, 1932.

Renny, C. *Medical Report on the Mahamurree in Gurhwal in 1849–50* (Agra, India: Secundra Orphan Press, 1851).

Revere, John. *An Inquiry into the Origins and Effects of Sulphurous Fumigations in the Cure of Rheumatism, Gout, Diseases of the Skin, Palsy &c*. Baltimore, MD: Coale & Co, 1822.

Richardson, M. W. "The rat and infantile paralysis: A theory." *Boston Medical and Surgical Journal* 175 (September 21, 1916): 397–400.

Ridlon, J. R. "Some aspects of ship fumigation." *Public Health Reports (1896–1970)* 46, no. 27 (July 3, 1931): 1572–1578.

Rivero, Pedro. "Saneamiento de la ciudad Buenos Aires. Deratización." *La Semana médica* 18, no.1 (1911): 1–31.

Roberts, Norman. "Cyanide fumigation of ships: Method used at New Orleans." *Public Health Reports (1896–1970)* 29, no. 50 (December 11, 1914): 3321–3325.

Rocher, Emile. *La province Chinoise du Yunnan.* Paris: Lerous, 1879.

Rodwell, James. *The Rat: Its History and Destructive Character.* London: Routledge and Co., 1858.

Sanders, James. "Report of the effects of the acid fumigation tried in Scotland during the prevalence of the epidemic cholera and the causes which prevented it from being every where known and adopted, &c." *London Medical and Surgical Journal* 3 (1833): 395–400.

Schnurrer, Friedrich. "Die geographische Verteilung der Krankheiten, vorgelesen in der Versammlung der deutschen Aerzte und Naturforscher zu München den 22. Sept. 1827." *Das Ausland* 1, no. März (1828): 357–59.

Schnurrer, Friedrich. *Die Cholera Morbus, ihre Verbreitung, ihre Zufälle, die versuchten Heilmethoden, ihre Eigenthümlichkeiten und die im Großen dagegen anzuwendenden Mittel.* Stuttgart, Tübingen: Gotta'sche Buchhandlung, 1831.

Scoutetten, H. *Relation historique et médicale de l'épidémie de choléra qui a régné à Berlin en 1831.* Paris: J.-B. Baillière, 1832.

Secques, Paul. *La dératisation par la chloropicrine.* Toulouse, France: Impr. P. Julia, 1937.

Serpossian, S. "La question des quarantaines, au sein de la Société Imperiale de Médecine." *Revue médico-pharmaceutique* 15, no. 6 (March 15, 1902): 67.

Serpossian, S. "La question des quarantaines, au sein de la Societe Imp. de Medicine." *Revue médico-pharmaceutique* 15, no. 7 (April 1, 1902): 76.

Service Sanitaire Maritime. "Sulfuration des navires.—Emploi d'appareils permettant d'effectuer cette sulfuration avant déchargement, circulaire du president du conseil, ministre de l'interieur et des cultes, du 20 juillet 1903, aux directeurs de la santé." *Recueil des Travaux du Comité Consultatif d'Hygiène Publique de France et des Actes Officiels de l'Administration Sanitaire* 33 (1903): 103–105.

Simond, Paul-Louis. "La propagation de la Peste." *Annales de l'Institut Pasteur* 62 (1898): 80–98.

Simpson, W. J. *A Treatise on Plague; Dealing with the Historical, Epidemiological, Clinical, Therapeutic and Preventive Aspects of the Disease.* Cambridge: Cambridge University Press, 1905.

Smith, C. "Germany. Method of killing rats at Hamburg." *Public Health Reports (1896–1970)* 15, no. 17 (April 27, 1900): 1012–1014.

Société Imperiale de Médecine. "Séance du 4 mai 1900; L'anhydride carbonique comme moyen de destruction des rats dans les cales des bateaux." *Gazette médicale d'Orient* 45, no. 7 (May 30, 1900): 108–113.

Société Imperiale de Médecine. "Séance du 18 mai 1900; L'anhydride carbonique comme moyen de destruction des rats dans les cales des bateaux, surtout en temps d'épidémie de peste." *Gazette médicale d'Orient* 45, no. 9 (June 30, 1900): 134–140.

Société Imperiale de Médecine. "Séance du 29 novembre 1901: Les quarantaines." *Gazette médicale d'Orient* 46, no. 22 (January 15, 1902): 886–891.

Société Imperiale de Médecine. "Séance du 13 decembre 1901, Les quarantaines." *Gazette médicale d'Orient* 46, no. 23 (February 1, 1902): 904–913.

Société Imperiale de Médecine. "Séance du 20 decembre 1901." *Gazette médicale d'Orient* 46, no. 24 (February 15, 1902): 915–917.

Stekoulis, C. "Bulletin épidémiologique." *Gazette médicale d'Orient* 42, no. 11 (July 31, 1897): 169–171.

Stekoulis, C. "Bulletin épidémiologique." *Gazette médicale d'Orient* 43, no. 24 (February 15, 1899): 353–354.

Stekoulis, C. "Bulletin épidémiologique." *Gazette médicale d'Orient* 45, no. 17 (October 31, 1900): 331–332.

Stekoulis, C. "Bulletin épidémiologique." *Gazette médicale d'Orient* 46, no. 6 (May 15, 1901),

Stekoulis, C. "Bulletin épidémiologique." *Gazette médicale d'Orient* 46, no. 12 (August 15, 1901): 721–723.

Stekoulis, C. "Bulletin épidémiologique." *Gazette médicale d'Orient* 46, no. 13 (September 1, 1901): 738–739.

Stekoulis, C. "La peste—La cholera." *Gazette médicale d'Orient* 44, nos. 23 and 24 (January 30 and February 15, 1900): 307–308.

Stekoulis, C. "Bulletin épidémiologique." *Gazette médicale d'Orient* 42, no. 11 (July 31, 1897): 169–171.

Swaine Taylor, Alfred. "An account of the Grotta del Cane; with remarks on suffocation by carbonic acid." *The London Medical and Physical Journal*, NS 73 (October 1832): 278–285.

Taptas, Nicholas. "Peste bubonique. Étude clinique et bactériologique." *Gazette médicale d'Orient* 42, no. 13 (August 31, 1897): 210–214.

Taptas, Nicholas. "Peste bubonique. Étude clinique et bactériologique." *Gazette médicale d'Orient* 42, no. 15 (September 15, 1897): 227–230.

Valassopoulo, Aristide. *La peste d'Alexandrie en 1899 au point de vue clinique, épidémiologique, etc.* Paris: A. Maloine, 1901.

Vallin, Émile Arthur. *Traité des désinfectants et de la désinfection.* Paris: G. Masson, 1882.

Vallin, Émile Arthur. "Rapport. Les services sanitaires et le lazaret du Frioul (séance 11 mars 1902)." *Bulletin de l'Académie nationale de médecine* 3, no. XLVII (1902): 311–341.

Walker, George Alfred. *A Treatise on the Cure of Ulcers by Fumigation.* London: Longman, 1847.

White, Charles Brainon. *Disinfection in Yellow Fever, as Practised at New Orleans, in the Years 1870 to 1875 Inclusive.* New Orleans: J. W. Madden, 1876.

Williams, C. L. "The air jet hydrocyanic acid sprayer." *Public Health Reports (1896–1970)* 46, no. 30 (July 24, 1931): 1755–1761.

Williams, C. L. "Notes on the fumigation of vessels. Preliminary inspection, how rats escape, increased periods of exposure, and other miscellaneous notes." *Public Health Reports (1896–1970)* 46, no. 50 (December 11, 1931): 2973–2980.

Williams, C. L. "The fumigation of loaded ships." *Public Health Reports (1896–1970)* 46, no. 31 (July 31, 1931): 1823–1836.

Wilm, Staff-Surgeon. *A Report on the Epidemic of Bubonic Plague at Hongkong in the Year 1896.* Hong Kong: n.p., 1897.

Woodworth, C.W. *School of Fumigation, University of California: Held at Pomona, California.* Los Angeles: Braun Corp., 1915.

Worell, Dorothy. *The Women's Municipal League of Boston. A History of Thirty-Five Years of Civic Endeavor.* Boston: Women's Municipal League of Committees Inc., 1943.

XXII Congrès international d'hygiène et de démographie. *Compte rendus du Congrès, Tome VII Premiere Division—Hygiène, Section VI, Hygiène administrative.* Brussels: P. Weissenbruch, 1903.

Yersin, Alexandre. "La peste bubonique à Hong Kong." *Annales de l'Institut Pasteur* 8 (1894): 662–667.

Zavitziano, Spiridion C. "Turkey. Report from Constantinople." *Public Health Reports (1896–1970)* 16, no. 12 (March 22, 1901): 602–604.

Zavitziano, Spiridion C. "Bulletin épidémiologique." *Gazette médicale d'Orient* 46, no. 14 (September 15, 1901): 753–754.

Zitterer, Charles. "Rapport sur les expériences de dératisation des cales des navires par le gaz acide carbonique présenté dans le Conseil Superieur de Santé (séance du 8/21 janvier 1902)." *Revue médico-pharmaceutique* 15, no. 1 (January 1, 1902): 14–16.

Zuschlag, Emil. *Le rat migratoire et sa destruction rationnelle*, trad. M. Pierre Oesterby. Copenhagen: Impr. F. Bagge, 1903.

Secondary Sources

Abensour, Miguel. "Persistent utopia." *Constellations* 15, no. 3 (2008): 406–421.

Ackerknecht, Erwin H. "Anticontagionism between 1821 and 1867." *Bulletin of the History of Medicine* 22 (1948): 562–93.

Almeida, Maria Antónia Pires de. "Epidemics in the news: Health and hygiene in the press in periods of crisis." *Public Understanding of Science* 22, no. 7 (February 12 2013): 886–902.

Alvarez, Adriana. "Resignificando los conceptos de la higiene: El surgimiento de una autoridad sanitaria en el Buenos Aires de los años 80." *História, Ciências, Saúde-Manguinhos* 6, no. 2 (1999): 293–314.

Alvarez, Adriana, and Adrián Carbonetti. *Saberes y prácticas médicas en la Argentina: Un recorrido por historias de vida*. Buenos Aires: EUDEM, 2008.

Amador, José. *Medicine and Nation Building in the Americas, 1890–1940*. Nashville, TN: Vanderbilt University Press, 2015.

Armus, Diego. "El descubrimiento de la enfermedad como problema social." In *El progreso, la modernización y sus límites (1880–1916)*, ed. Mirta Zaida Lobato, 507–552. Buenos Aires: Editorial Sudamericana, 2002

Armus, Diego. "Disease in the historiography of modern Latin America." In *Disease in the History of Modern Latin America: From Malaria to AIDS*, ed. Diego Armus, 1–24. Durham, NC: Duke University Press, 2003.

Armus, Diego. *The Ailing City: Health, Tuberculosis, and Culture in Buenos Aires, 1870–1950*. Durham, NC: Duke University Press, 2011.

Armus, Diego. "Utopías higienicas/utopías urbanas. Buenos Aires 1920." In *Utopias urbanas: Geopoliticos del deseo en América Latina*, ed. Gisela Heffes, 115–130. Frankfurt/Madrid: Verveurt, 2013.

Arslan, A., and H. A. Polat. "Travel from Europe to Istanbul in the 19th century and the quarantine of Çanakkale." *Journal of Transport & Health* (2017): 10–17.

Au, Sokhieng. *Mixed Medicines: Health and Culture in French Colonial Cambodia*. Chicago, IL: The University of Chicago Press, 2011.

Baldwin, Peter. *Contagion and the State in Europe, 1830–1930.* Cambridge: Cambridge University Press, 2005.

Barnes, Donna A. *The Louisiana Populist Movement, 1881–1900.* Baton Rouge: Louisiana State University Press, 2011.

Barnes, David S. "'Until cleansed and purified': Landscapes of health in the interpermeable world." *Change Over Time* 6, no. 2 (November 10 2016): 138–152.

Barnes, David S. "Cargo, 'infection,' and the logic of quarantine in the nineteenth century." *Bulletin of the History of Medicine* 88, no. 1 (2014): 75–101.

Bashford, Alison. *Imperial Hygiene: A Critical History of Colonialism, Nationalism and Public Health.* Basingstoke, UK: Palgrave Macmillan, 2004.

Bashford, Alison. "Maritime quarantine: Linking Old World and New World histories." In *Quarantine: Local and Global Histories*, ed. Alison Bashford, 1–12. London: Palgrave Macmillan, 2016.

Bashford, Alison (ed.). *Quarantine: Local and Global Histories.* London: Palgrave Macmillan, 2016.

Bein, Amit. "The Istanbul earthquake of 1894 and science in the late Ottoman Empire." *Middle Eastern Studies* 44, no. 6 (2008): 909–924.

Belmartino, Susana. *La atención médica argentina en el siglo XX: Instituciones y procesos.* Siglo Veintiuno Editores Argentina, 2005.

Belmar, Antonio García, and Ramón Bertomeu-Sánchez. "L'Espagne fumigée. Consensus et silences autour des fumigations d'acides minéraux en Espagne, 1770–1804." *Annales historiques de la Révolution française* 383, no. 1 (2016): 177–202.

Benchimol, Jaime Larry. *Dos micróbios aos mosquitos: Febre amarela e a revolução pasteuriana no Brasil.* Rio de Janeiro: SciELO—Editora FIOCRUZ, 1999.

Biehler, Dawn Day. *Pests in the City: Flies, Bedbugs, Cockroaches, and Rats.* Washington, DC: Washington University Press, 2013.

Blancou, J. "History of disinfection from early times until the end of the 18th century." *Revue scientifique et technique* (International Office of Epizootics) 14, no. 1 (March 1995): 21–39.

Bloch, Ernst. *The Principle of Hope*, 3 volumes, trans. by Neville Plaice, Stephen Plaice, and Paul Knight. Cambridge, MA: MIT Press, 1995.

Bluma, Lars. "The hygienic movement and German mining 1890–1914." In *History of the Workplace: Environment and Health at Stake*, ed. Lars Bluma and Judith Rainhorn, 7–26. London and New York: Routledge, 2015.

Bouguer, Pierre. *Traité du navire, de sa construction, et de ses mouvemens*. Paris: Jombert, 1746.

Bradley, Mark (ed.). *Rome, Pollution and Propriety. Dirt, Disease and Hygiene in the Eternal City from Antiquity to Modernity*. Cambridge: Cambridge University Press, 2012.

Brandt, Allan M. *No Magic Bullet: A Social History of Venereal Disease in the United States since 1880*. New York: Oxford University Press, 1987.

Broemer, R. "The first global map of the distribution of human diseases: Friedrich Schnurrer's 'Charte über die geographische Ausbreitung der Krankheiten' (1827)." *Medical History* Supplement 20 (2000): 176–185.

Brown, Hannah, and Ann Kelly. "Material proximities and hotspots: Toward an anthropology of viral hemorrhagic fevers." *Medical Anthropology Quarterly* 28, no. 2 (2014): 280–303

Bruzzone, Rachel. "*Polemos, pathemata*, and plague: Thucydides' narrative and the tradition of upheaval." *Greek, Roman, and Byzantine Studies* 57 (2017): 882–909.

Bulmus, Birsen. *Plague, Quarantines and Geopolitics in the Ottoman Empire*. Edinburgh: Edinburgh University Press, 2012.

Burgess, Douglas. *Engines of Empire: Steamships and the Victorian Imagination*. Stanford, CA: Stanford University Press, 2017.

Byrne, Joseph P. *Encyclopedia of the Black Death*. Santa Barbara, CA: ABC-CLIO, 2012.

Campanella, Richard. "The battle against bubonic plague: 100 years ago, New Orleans waged war on rats." *The Times Picayune*, May 8, 2014. http://www.nola.com/homegarden/index.ssf/2014./08/one_hundred_years_ago_as_war_b.html

Chahrour, Marcel. "'A civilizing mission'? Austrian medicine and the reform of medical structures in the Ottoman Empire, 1838–1850." *Studies in History and Philosophy of Biological and Biomedical Sciences* 38, no. 4 (December 2007): 687–705.

Chang, Hasok. "The persistence of epistemic objects through scientific change." *Erkenntnis* 75, no. 3 (November 1 2011): 413–429.

Chevallier, Jacques. "Une quarantaine de peste au lazaret de Frioul en 1901." *Histoire des sciences médicales* XLIX, no. 2 (2015): 179–188.

Chiffoleau, Sylvia. "Les pèlerins de La Mecque, les germes et la communauté internationale." *Médecine/sciences (Paris)* 27, no. 12 (December 2011): 1121–1126.

Coleman, William. *Yellow Fever in the North. The Methods of Early Epidemiology*. Madison: University of Wisconsin Press, 1987.

Collier, Stephen J., Jamie Cross, Peter Redfield, and Alice Street (eds.), "Little development devices/humanitarian goods," *Limn* 9 (November 2017), https://limn.it/issues/little-development-devices-humanitarian-goods/.

Connor, Steven. *The Matter of Air: Science and Art of the Ethereal.* London: Reaktion Books, 2010.

Conrad, Sebastian. *Globalisation and the Nation in Imperial Germany.* Cambridge: Cambridge University Press, 2010.

Crary, Jonathan. *Techniques of the Observer. On Vision and Modernity in the Nineteenth Century.* Cambridge, MA: MIT Press, 1990.

Cueto, Marcos. *The Value of Health: A History of the Pan American Health Organization.* Washington, DC: Pan American Health Organization, 2007.

Cueto, Marcos, and Steven Palmer. *Medicine and Public Health in Latin America.* Cambridge: Cambridge University Press, 2014.

Debus, Allen G. *The Chemical Philosophy.* Mineola, NY: Dover Publications, 1977.

Debus, Allen G. *The English Paracelsians: The Chemical Challenge to Medical and Scientific Tradition in Early Modern France.* Cambridge: Cambridge University Press, 1991.

de Laet, Marianne and Annemarie Mol. "The Zimbabwe bush pump: Mechanics of a fluid technology." *Social Studies of Science* 30, no. 2 (April 2000): 225–263.

Delaporte, François. *The History of Yellow Fever: An Essay on the Birth of Tropical Medicine,* trans. Arthur Goldhammer. Cambridge, MA: MIT Press, 1991.

Deleuze, Gilles, and Felix Guattari. *Capitalism and Schizophrenia I: Anti-Oedipus,* trans. Robert Hurley. London: Penguin, 2009.

della Dora, Veronica. "Making mobile knowledges: The educational cruises of the Revue générale des sciences pures et appliquées, 1897–1914." *Isis* 101, no. 3 (September 2010): 467–500.

Demir, Alpaslan. "Mice problems in the Ottoman Empire and mice invasion in Tirhala in 1866." *IBAC* 2 (2012): 645–661.

de Morsier, Georges. "La vie et l'œuvre de Louis Odier, docteur et professeur en médecine (1748—1817)." *Gesnerus* 32 (1975): 248–270.

Déré, Anne-Claire. "Daniel August Rosenstiehl (1839–1916): An Alsatian chemist in the synthetic dyestuffs industry." In *The Chemical Industry in Europe, 1850–1914: Industrial Growth, Pollution, and Professionalization,* ed. Erns Homburg, Anthony S. Travis, and Harm G. Schröter, 305–320. Dordrecht, Germany: Springer, 1998.

Douglas, Mary. *Purity and Danger, an Analysis of Concepts of Pollution and Taboo.* London and New York, Routledge, 2002 [1966].

Echenberg, Myron. *Plague Ports: The Global Urban Impact of Bubonic Plague, 1894–1901.* New York: New York University Press, 2007.

Eggert, Katherine. *Disknowledge: Literature, Alchemy, and the End of Humanism in Renaissance England.* Philadelphia: University of Pennsylvania Press, 2015.

Erer, S., and A. D. Erdemir, "Preventive measures and treatments for cholera in the 19th century in Ottoman archive documents." *Vesalius* 16, no. 1 (June 2010): 41–48.

Ersoy, Nermin, Yuksel Gungor, and Aslihan Akpinar. "International Sanitary Conferences from the Ottoman perspective (1851–1938)." *Hygiea Internationalis: An Interdisciplinary Journal for the History of Public Health* 10, no. 1 (2011): 53–79.

Evans, Nicholas. "'Blaming the rat? Accounting for plague in colonial Indian medicine." *Medicine, Anthropology, Theory* 5, no. 3 (2018) doi.org/10.17157/mat.5.3.371

Evans, Richard. *Death in Hamburg: Society and Politics in the Cholera Years, 1830–1910.* Wotton-under-Edge, UK: Clarendon Press, 1987.

Faber, Knut. *Nosography. The Evolution of Clinical Medicine in Modern Times.* New York: AMS, 1930.

Faroqhi, Suraiya (ed.). *Animals and People in the Ottoman Empire.* Istanbul: Eren, 2010.

Fitzharris, Lindsey. *The Butchering Art: Joseph Lister's Quest to Transform the Grisly World of Victorian Medicine.* London: Penguin, 2017.

Fleischer, Bernhard. "A century of research in tropical medicine in Hamburg: the early history and present state of the Bernhard Nocht Institute." *Tropical Medicine and International Health* 5, no. 10 (2000): 747–751.

Foucault, Michel, and Jay Miskowiec. "Of Other Spaces." *Diacritics* 16, no. 1 (1986): 22–27. https://doi.org/10.2307/464648.

Garber, James J. *Harmony in Healing: The Theoretical Basis of Ancient and Medieval Medicine.* London and New York: Routledge, 2008.

Gaillard, Béatrice. "Les Franque et les bâtiments hospitaliers d'Avignon au XVIIIe siècle: entre tradition et mutations." *In Situ: Revue des patrimonies* 31 (2017). http://insitu.revues.org/14242.

Geissler, Wenzel Paul (ed.). *Para-states and Medical Science: Making African Global Health.* Durham, NC: Duke University Press, 2015.

Gillson, Gordon. *Louisiana State Board of Health: The Formative Years.* Baton Rouge: Louisiana State Board of Health, 1967.

Gillson, Gordon. *Louisiana State Board of Health: The Progressive Years.* Baton Rouge: Louisiana State Board of Health, 1976.

Gradmann, Christoph. "A spirit of scientific rigour: Koch's postulates in twentieth-century medicine." *Microbes and Infection* 16, no. 11 (November 2014): 885–892.

Gürsel, Zeynep Devrim. "Thinking with X-rays: Investigating the politics of visibility through the Ottoman Sultan Abdülhamid's photography collection." *Visual Anthropology* 29, no. 3, Special Issue: Visual Revolutions in the Middle East (2016): 229–242.

Haber, L. F. *The Poisonous Cloud: Chemical Warfare in the First World War*. Oxford: Oxford University Press, 1986.

Hardy, Anne. *The Epidemic Streets: Infectious Disease and the Rise of Preventive Medicine, 1856–1900*. Wotton-under-Edge, UK: Clarendon Press, 1993.

Harrison, Mark. *Contagion: How Commerce Has Spread Disease*. New Haven, CT: Yale University Press, 2013.

Haycock, David Boyd, and Sally Archer (eds). *Health and Medicine at Sea, 1700–1900*. Woodbridge: Boydell Press, 2009.

Haynes, Williams. *Brimstone, the Stone That Burns: The Story of the Frasch Sulphur Industry*. Princeton, NJ: Van Nostrand, 1959.

Heiningen, Teunis Willem van. "La contribution à la santé publique de Louis-Bernard Guyton de Morveau (1737–1816) et l'adoption de ses idées aux Pays-Bas." *Histoire des sciences médicales* XLVIII, no. 1 (2014): 97–106.

Henderson, John. "Filth is the mother of corruption: Plague and the built environment in early modern Florence." In *Plague and the City*, eds. Lukas Engelmann, John Henderson, and Christos Lynteris, 69–90. London and New York: Routledge, 2018.

Honigsbaum, Mark. "'Tipping the balance': Karl Friedrich Meyer, latent infections, and the birth of modern ideas of disease ecology." *Journal of the History of Biology* 49, no. 2 (2016): 261–309.

Howard-Jones, Norman, and the World Health Organization. *The Scientific Background of the International Sanitary Conferences, 1851–1938*. Geneva: World Health Organization, 1975.

Humphreys, Margaret. *Yellow Fever and the South*. Baltimore, MD: Johns Hopkins University Press, 1999.

Jasanoff, S. and S.-H. Kim (eds). *Dreamscapes of Modernity: Sociotechnical Imaginaries and the Fabrication of Power*. Chicago, IL: University of Chicago Press, 2015.

Jouanna, Jacques. *Greek Medicine from Hippocrates to Galen: Selected Papers*. Leiden, Netherlands: Brill, 2012.

Kalimtzis, Kostas. *Aristotle on Political Enmity and Disease. An Inquiry into Stasis*. New York: SUNY Press, 2000.

Karatzanis, Alexander, Constantinos Trompoukis, Ioannis Vlastos, and George Velegrakis. "On the history of modern tonsillectomy: The contribution of Nikolaos Taptas." *European Archives of Oto-Rhino-Laryngology* 268, no. 1687 (November 2011). https://link.springer.com/article/10.1007/s00405-011-1751-3.

Kelly, Catherine. "Parliamentary inquiries and the construction of medical argument in the early 19th century, 1793–1825." In *Lawyers' Medicine: The Legislature, the Courts and Medical Practice, 1760–2000*, eds. Imogen Goold and Catherine Kelly, 17–38. London: Bloomsbury, 2000.

King, Helen. *Hippocrates' Woman: Reading the Female Body in Ancient Greece*. London and New York: Routledge, 1998.

Koch, Tom. *Disease Maps: Epidemics on the Ground*. Chicago, IL: University of Chicago Press, 2011.

Kohl, Alejandro. *Higienismo Argentino: Historia de una utopía. La salud en el imaginario colectivo de una época*. Buenos Aires: Editorial Dunken, 2006.

Koselleck, Reinhart. *The Practice of Conceptual History: Timing History, Spacing Concepts*. Stanford, CA: Stanford University Press, 2002.

Kraut, Alan M. *Silent Travelers. Germs, Genes, and the "Immigrant Menace."* New York: BasicBooks, 1995.

Kroner, Hans-Peter. "From eugenics to genetic screening. Historical problems of human genetic applications." In *The Ethics of Genetic Screening*, eds. Ruth F. Chadwick, Darren Shickle, H.A. Ten Have, and Urban Wiesing, 131–140. Berlin: Springer Science 1999.

Kuhnke, LaVerne. *Lives at Risk: Public Health in Nineteenth-Century Egypt*. Berkeley, CA: University of California Press, 1990.

Lacorte, J. Guilherme. "A atuação de Oswaldo Cruz no aparecimento da peste bubônica no Brasil." *A Folha médica* 54 (1967): 183–188.

Latour, Bruno. *The Pasteurization of France*, trans. Alan Sheridan. Cambridge, MA: Harvard University Press, 1988.

Levitas, Ruth. "Educated hope: Ernst Bloch on abstract and concrete utopia." *Utopian Studies* 1, no. 2 (1990): 13–26.

Levitas, Ruth. *The Concept of Utopia*. Oxford: Peter Land, 2010.

Liscia, María Silvia Di. "Del brazo civilizador a la defensa nacional: Políticas sanitarias, atención médica y población rural (Argentina, 1900–1930)." *Historia Caribe* 12, no. 31 (2017): 159–93.

Lorey, Isabell. *Figuren Des Immunen. Elemente Einer Politischen Theorie*. Zürich: Diaphanes, 2011.

Low, Michael Christopher. *The Mechanics of Mecca: The Technopolitics of the Late Ottoman Hijaz and the Colonial Hajj.* PhD Thesis, Columbia University, 2015.

Lynteris, Christos. "'A suitable soil': Plague's breeding grounds at the dawn of the third pandemic." *Medical History* 61, no. 3 (June 2017): 343–357.

Lynteris, Christos. "Vagabond microbes, leaky labs and epidemic mapping: Alexandre Yersin and the 1898 plague epidemic in Nha Trang." *Social History of Medicine* (2019), https://doi.org/10.1093/shm/hkz053.

Lynteris, Christos, and Nicholas H. Evans (eds). *Histories of Post-Mortem Contagion: Infectious Corpses and Contested Burials.* London: Palgrave Macmillan, 2017.

Lynteris, Christos, and Branwyn Poleykett. "The anthropology of epidemic control: Technologies and materialities." *Medical Anthropology* 37, no. 6 (2018): 433–441.

Maglen, Krista. *The English System: Quarantine, Immigration and the Making of a Port Sanitary Zone.* Manchester: Manchester University Press, 2014.

Markel, Howard. *Quarantine!: East European Jewish Immigrants and the New York City Epidemics of 1892.* Baltimore MD: Johns Hopkins University Press, 1999.

McLean, David. "Protecting wood and killing germs: 'Burnett's Liquid' and the origins of the preservative and disinfectant industries in early Victorian Britain." *Business History* 52, no. 2 (April 2010): 285–305.

Meik, Kindon Thomas. "Disease and hygiene in the construction of a nation: The public sphere, public space, and the private domain in Buenos Aires, 1871–1910." Ph.D. Thesis, Florida International University, 2011.

Meinel, Fabian. *Pollution and Crisis in Greek Tragedy.* Cambridge: Cambridge University Press, 2015.

Mendelsohn, J. Andrew. "From eradication to equilibrium. How epidemics became complex after World War I." In *Greater than the Parts: Holism in Biomedicine, 1920–1950,* eds. Christopher Lawrence and George Weisz, 303–334. Oxford: Oxford University Press, 1998.

Mendelsohn, J. Andrew. "'Like all that lives': Biology, medicine and bacteria in the age of Pasteur and Koch." *History and Philosophy of the Life Sciences* 24, no. 1 (2002): 3–36.

Mikhail, Alan. *The Animal in Ottoman Egypt.* Oxford: Oxford University Press, 2014.

Mishra, Saurab. "Incarceration and resistance in a Red Sea lazaretto, 1880–1930." In *Quarantine: Local and Global Histories,* ed. Alison Bashford, 54–65. London: Palgrave Macmillan, 2016.

Mishra, Saurab. *Pilgrimage, Politics, and Pestilence: The Haj from the Indian Subcontinent.* Oxford: Oxford University Press, 2011.

Mitchell-Boyask, Robin. *Plague and the Athenian Imagination: Drama, History, and the Cult of Asclepius*. Cambridge: Cambridge University Press, 2007.

Miyazaki, Hirokazu. *The Method of Hope: Anthropology, Philosophy, and Fijian Knowledge*. Stanford, CA: Stanford University Press, 2004.

Moir, Cat. "Casting a picture. Utopia, Heimat and the materialist concept of history." *Anthropology and Materialism* 3 (2016). https://am.revues.org/573.

Monmonier, Mark. "Maps as graphic propaganda for public health." In *Imagining Illness: Public Health and Visual Culture*, ed. David Harley Serlin, 108–25. Minneapolis: University of Minnesota Press, 2010.

Montesu, M. A., and F. Cottoni. "G. C. Bonomo and D. Cestoni. Discoverers of the parasitic origin of scabies." *American Journal of Dermatopathology* 13, no. 4 (August 1991): 425–427.

Mooney, Graham. *Intrusive Interventions: Public Health, Domestic Space, and Infectious Disease Surveillance in England, 1840–1914*. Rochester, NY: University of Rochester Press, 2015.

Moulin, Anne-Marie. "Initiating global health at the time of the Crimean war 1853–1856 and the projects of sanitary reform of the Ottoman Empire." *History of Medicine* 1 (2014): 61–79.

Napier, David. *The Age of Immunology: Conceiving a Future in an Alienating World*. Chicago, IL: The University of Chicago Press, 2003.

Nash, Linda. *Inescapable Ecologies: A History of Environment, Disease, and Knowledge*. Berkeley, CA: The University of California Press, 2006.

Newman, William R. "Alchemical and chymical principles. Four different traditions." In *The Idea of Principles in Early Modern Thought: Interdisciplinary Perspectives*, ed. Peter R. Anstey, 77–97. London and New York: Routledge, 2017.

Nicolas, Michele. "Pierre Apéry et ses publications scientifiques." *Revue d'histoire de la pharmacie* 94, no. 350 (2006): 237–247.

Olivarius, Kathryn. "Immunity, capital, and power in antebellum New Orleans." *American Historical Review* 124, no. 2 (2019): 425–455.

Olseen, Mark. "Totalitarianism and the 'repressed' utopia of the present: Moving beyond Hayek, Popper and Foucault." *Policy Futures in Education* 1, no. 3 (2003): 526–552.

O'Rourke, Kevin H., and Jeffrey G. Williamson. *Globalization and History: The Evolution of a Nineteenth-Century Atlantic Economy*. Cambridge, MA: MIT Press, 1999.

Packhard, Randall. *A History of Global Health: Interventions into the Lives of Other Peoples*. Baltimore, MD: Johns Hopkins University Press, 2016.

Panzac, D. *Quarantaines et lazarets: L'Europe et la peste d'Orient (XVII°-XX° siècle)*. Aix-en-Provence, France: Edisud, 1986.

Panzac, D. *La peste dans l'Empire ottoman: 1700–1850*. Leuven, Belgium: Editions Peeters, 1985.

Pemberton, Neil. "The rat-catcher's prank: Interspecies cunningness and scavenging in Henry Mayhew's London." *Journal of Victorian Culture*, 19, no. 4 (2014): 520–535.

Pietsch, Tamson. "A British sea: making sense of global space in the late nineteenth century." *Journal of Global History* 5, no. 3 (November 2010): 423–446.

Pinault, Jody Rubin. *Hippocratic Lives and Legends*. Leiden, Netherlands: Brill, 1992.

Poleykett, Branwyn. "Building out the rat: animal intimacies and prophylactic settlement in 1920s South Africa." *Engagement*, February 2, 2017. https://aesengagement
.wordpress.com/2017/02/07/building-out-the-rat-animal-intimacies-and-prophylactic
-ssettlement-in-1920s-south-africa/.

Ramos-e-Silva, Marcia. "Giovan Cosimo Bonomo (1663–1696): Discoverer of the etiology of scabies." *International Journal of Dermatology* 37, no. 8 (August 1998): 625–630.

Rasimoğlu, Ceren Gülser İlikan. "The foundation of a professional group: Physicians in the nineteenth century modernizing century modernizing Ottoman Empire (1839–1908)." Ph.D. Thesis, Boğaziçi University, 2012.

Rawcliffe, Carole. "'Great stenches, horrible sights and deadly abominations': Butchery and the battle against plague in late medieval English towns." In *Plague and the City*, ed. Lukas Engelmann, John Henderson, and Christos Lynteris, 18–38. London and New York: Routledge, 2018.

Rebelo, Fernanda. "Between the Carlo R. and the Orleannais: Public health and maritime prophylaxis in the description of two cases of ships transporting immigrants arriving in the port of Rio de Janeiro, 1893–1907." *História, Ciências, Saúde-Manguinhos* 20, no. 3 (September 2013): 765–796.

Redfield, Peter. "Fluid technologies: The bush pump, the LifeStraw® and microworlds of humanitarian design." *Social Studies of Science* 46, no. 2 (April 2016): 159–183.

Reubi, David. "A genealogy of epidemiological reason: Saving lives, social surveys and global population." *BioSocieties* 13, no. 1 (March 1 2018): 81–102.

Rheinberger, Hans-Jörg. "Experimental systems. Historiality, narration, and deconstruction." In *The Science Studies Reader*, ed. Mario Biagioli, 417–429. London and New York: Routledge, 1999.

Risse, Guenter B. *Driven by Fear: Epidemics and Isolation in San Francisco's House of Pestilence*. Champaign, IL: University of Illinois Press, 2015.

Roark, R. C. *A Bibliography of Chloropicrin 1848–1932*. Washington, DC: US Department of Agriculture, Miscellaneous Publication No. 176, 1934.

Rodríguez, Julia. *Civilizing Argentina: Science, Medicine, and the Modern State*. Chapel Hill, NC: University of North Carolina Press, 2006.

Rogaski, Ruth. *Hygienic Modernity. Meanings of health and Disease in Treaty-Port China*. Berkeley, CA: The University of California Press, 2004.

Rosenberg, Charles E. "The therapeutic revolution: Medicine, meaning and social change in nineteenth-century America." *Perspectives in Biology and Medicine* 20, no. 4 (1977): 485–506.

Ruggiero, Kristin. *Modernity in the Flesh: Medicine, Law, and Society in Turn-of-the-Century Argentina*. Stanford, CA: Stanford University Press, 2004.

Russell, Edmund. *War and Nature: Fighting Humans and Insects with Chemicals from World War I to* Silent Spring. Cambridge: Cambridge University Press, 2001.

Sánchez, Norma Isabel, and Alfredo G. Kohn Loncarica. *La higiene y los higienistas en la Argentina: 1880–1943*. Buenos Aires: Sociedad Científica Argentina, 2007.

Sari, N. "Turkey and its international relations in the history of medicine." *Vesalius* VII, no. 2 (2001): 86–93.

Sariyildiz, G., and O. D. Macar, "Cholera, pilgrimage, and international politics of sanitation: The quarantine station on the island of Kamaran." In *Plague and Contagion in the Islamic Mediterranean*, ed. Nukhet Varlik, 243–274. Croydon, UK: ARC Humanities Press, 2017.

Schaffer, Simon. "Measuring virtue: Eudiometry, enlightenment, and pneumatic medicine." In *The Medical Enlightenment of the Eighteenth Century*, ed. Andrew Cunningham and Roger French, 281–318. Cambridge: Cambridge University Press, 1990.

Schaffer, Simon, David Serlin, and Jennifer Tucker. "Editors' introduction. Special Issue : The history of technoscience." *Radical History Review* 127 (January 1, 2017): 1–12.

Schlich, Thomas. "Asepsis and bacteriology: A realignment of surgery and laboratory science." *Medical History* 56, no. 3 (July 2012): 308–334.

Serrano, Elana. "Spreading the evolution: Guyton's fumigating machine in Spain. Politics, technology, and material culture (1796–1808)." In *"Astonishing Transformations": How Chemistry Made and Managed the World, 1760–1840*, ed. Lissa Roberts and Simon Warrett, 106–130. Amsterdam: Brill, 2017.

Shah, Nayan. *Contagious Divides: Epidemics and Race in San Francisco's Chinatown*. Berkeley, CA: University of California Press, 2001.

Shapin, Steven, and Simon Schaffer. *Leviathan and the Air-Pump: Hobbes, Boyle, and the Experimental Life*. Princeton, NJ: Princeton University Press, 2011.

Shefer-Mossensohn, Miri. *Science among the Ottomans: The Cultural Creation and Exchange of Knowledge*. Austin: University of Texas Press, 2015.

Simond, M. "Paul-Louis Simond and his discovery of plague transmission by rat fleas: A centenary." *Journal of the Royal Society of Medicine* 91 (2) (February 1998): 101–104.

Sturdy, S. W. "A co-ordinated whole: The life and work of John Scott Haldane." Ph.D. Thesis, University of Edinburgh, 1987. https://www.era.lib.ed.ac.uk/handle/1842/6873.

Tekiner, Halil, and Afife Mat. "Les pharmacopées turques de langue française." *Revue d'histoire de la pharmacie* 57, no. 361 (2009): 17–22.

Temkin, Owsei. "An historical analysis of the concept of infection." *Studies in Intellectual History* (1953): 123–147.

Thompson, E. P. *William Morris: Romantic or Revolutionary?* 2nd revised edition. London: The Merlin Press, 2011.

Tilles, Gérard, and Daniel Wallach. "Le traitement de la syphilis par le mercure. Une histoire thérapeutique exemplaire." *Revue d'histoire de la pharmacie* 84, no. 312 (1996): 347–351.

Trouillot, Michel-Rolph. *Silencing the Past: Power and the Production of History*. Boston, MA: Beacon Press, 2015.

Vann, Michael G. "Of rats, rice, and race: The great Hanoi rat massacre, an episode in French colonial history." *French Colonial History* 4 (2003): 191–203.

Varlık, Nükhet. *Plague and Empire in the Early Modern Mediterranean World: The Ottoman Experience, 1347–1600*. Cambridge: Cambridge University Press, 2015.

von Ezdorf, R. H. "The occurrence of plague in Habana and the measures adopted for its control and eradication." *Public Health Reports (1896–1970)* 27, no. 42 (October 18, 1912): 1697–1702.

Weindling, Paul. "The uses and abuses of biological technologies: Zyklon B and gas disinfestation between the First World War and the Holocaust." *History and Technology* 11, no. 2 (1994): 291–298.

Whyte, Rebecca. "Disinfection in the laboratory: Theory and practice in disinfection policy in late C19th and Early C20th England." *Endeavour* 39, no. 1 (March 2015): 35–43.

Worboys, Michael. *Spreading Germs: Disease Theories and Medical Practice in Britain, 1865–1900*. Cambridge: Cambridge University Press, 2000.

Yalçinkaya, M. Alper. *Learned Patriots: Debating Science, State, and Society in the Nineteenth-Century Ottoman Empire.* Chicago, IL: The University of Chicago Press, 2015.

Yıldırım, Nuran. *A History of Healthcare in Istanbul: Health Organizations—Epidemics, Infections and Disease Control Preventive Health Institutions—Hospitals—Medical Education.* Istanbul: Ajansfa, 2010.

Yıldırım, Nuran, and Hakan Ertin. "European physicians/specialists during the cholera epidemic in Istanbul 1893–1895 and their contributions to the modernization of healthcare in the Ottoman State." In *Health Culture and the Human Body: Epidemiology, Ethics and History of Medicine; Perspectives from Turkey and Central Europe*, ed. Ilhan Likilic, Hakan Ertin, Rainer Bromer, and Hajo Zeeb, 189–215. Istanbul: BETIM Center Press, 2014.

Index

Biopower, 25
Birdfoot Delta (US), 1
Black Death, 19, 32
Blackpool (UK), 133
Black Sea, 85, 120
Bloch, Ernst, 10
Boards of health
  Argentina, 152, 160
  Istanbul, 96, 114
  Louisiana State Board of Health, 4, 19,
    55, 57, 58, 63, 64, 65, 68, 71, 75, 76–
    78, 105, 116
Bonkowski Pasha, 95
Bordas (physician), 189
Bordeaux, 116
Borel, Frédéric, 128, 129
Boulton Bliss & Dallett Company, 185
Britain, 36, 41, 51, 66, 100, 106, 109.
    See also Quarantine Act (UK)
  British Board of Trade, 120
  Colonial Office, 106, 180
  House of Commons, 39
  UK Chamber of Shipping, 120
  British India Steam Navigation
    Company, 106
Bromide. *See under* Fumigation gas
Bronchitis, 133
Brooklyn, 67
Brown, Hannah, 178
Brussels Medical Congress, 128–129,
  130, 131, 181
Bucquoy, Jules, 110
Buenos Aires, 21, 146, 147, 244n1
Bunsen burner, 166, 179
Burgess, Douglas, 12
Butler, Benjamin, 60

Calcasieu Parish (Louisiana), 62
Calcutta, 113, 131, 180
Callao (Peru), 144
Calmette, Albert, 114–116, 129, 130,
  134, 139–141, 200
Cambodia, 145

Canary islands, 190
Cape of Good Hope, 179
Cape Town, 234n16
Capitalism, 12, 18
Carbolic acid/phenic acid, 50, 61–62,
  64, 65, 66, 67, 77, 94, 111, 120, 191,
  222n37. *See also* Phenol
Carbon dioxide. *See under* Fumigation
  gas
Carbon monoxide/carbonic oxide. *See
  under* Fumigation gas
Cargo, 2, 7, 62, 63, 70, 73, 92, 98, 100,
  102, 106, 107, 109, 111, 113, 116–
  119, 123, 126–142, 145, 160, 162,
  178, 182, 183, 186, 187, 190, 191,
  199, 202
Carroll, James, 6, 208n9
Cartagena (Spain), 40, 41
Cats, 184
Cattle plague, 50
Ceylon (Sri Lanka), 193–194
Charbonniere (quack doctor), 24
Charcoal, 28, 34, 35, 99, 101, 111, 120
Chemical warfare, 189, 204
China, 86, 145
Chlorine, 49, 52. *See also under*
  Fumigation gas
Chlorocyanic gas. *See under* Fumigation
  gas
Chloropicrin (CCl$_3$NO$_2$). *See under*
  Fumigation gas
Cholera, 2, 16, 41–46, 50, 52, 60, 63, 72,
  85, 87, 97, 98, 110, 114, 116, 126,
  127, 128, 134, 138, 148, 149, 151,
  161, 164, 167, 220n120
  fowl, 68
Cinnabar, 24, 33
Civil War (US), 60
Clairborne, William C., 56
Clayton, Thomas Adam, 55, 71, 73, 76,
  78
Clayton apparatus/machine. *See under*
  Fumigation apparatus

Inside Technology Series

Edited by Wiebe E. Bijker, W. Bernard Carlson, and Trevor J. Pinch

David Demortain, *The Science of Bureaucracy: Risk Decision-Making and the US Environmental Protection Agency*

Joeri Bruynincx, *Listening in the Field: Recording and the Science of Birdsong*

Edward Jones-Imhotep, *The Unreliable Nation: Hostile Nature and Technological Failure in the Cold War*

Jennifer L. Lieberman, *Power Lines: Electricity in American Life and Letters, 1882–1952*

Jess Bier, *Mapping Israel, Mapping Palestine: Occupied Landscapes of International Technoscience*

Benoît Godin, *Models of Innovation: The History of an Idea*

Stephen Hilgartner, *Reordering Life: Knowledge and Control in the Genomics Revolution*

Brice Laurent, *Democratic Experiments: Problematizing Nanotechnology and Democracy in Europe and the United States*

Cyrus C. M. Mody, *The Long Arm of Moore's Law: Microelectronics and American Science*

Tiago Saraiva, *Fascist Pigs: Technoscientific Organisms and the History of Fascism*

Teun Zuiderent-Jerak, *Situated Interventions: Sociological Experiments in Healthcare*

Basile Zimmermann, *Technology and Cultural Difference: Electronic Music Devices, Social Networking Sites, and Computer Encodings in Contemporary China*

Andrew J. Nelson, *The Sound of Innovation: Stanford and the Computer Music Revolution*

Sonja D. Schmid, *Producing Power: The Pre-Chernobyl History of the Soviet Nuclear Industry*

Casey O'Donnell, *Developer's Dilemma: The Secret World of Videogame Creators*

Christina Dunbar-Hester, *Low Power to the People: Pirates, Protest, and Politics in FM Radio Activism*

Eden Medina, Ivan da Costa Marques, and Christina Holmes, editors, *Beyond Imported Magic: Essays on Science, Technology, and Society in Latin America*

Anique Hommels, Jessica Mesman, and Wiebe E. Bijker, editors, *Vulnerability in Technological Cultures: New Directions in Research and Governance*

Amit Prasad, *Imperial Technoscience: Transnational Histories of MRI in the United States, Britain, and India*

Charis Thompson, *Good Science: The Ethical Choreography of Stem Cell Research*

Tarleton Gillespie, Pablo J. Boczkowski, and Kirsten A. Foot, editors, *Media Technologies: Essays on Communication, Materiality, and Society*

Catelijne Coopmans, Janet Vertesi, Michael Lynch, and Steve Woolgar, editors, *Representation in Scientific Practice Revisited*

Rebecca Slayton, *Arguments that Count: Physics, Computing, and Missile Defense, 1949–2012*

Stathis Arapostathis and Graeme Gooday, *Patently Contestable: Electrical Technologies and Inventor Identities on Trial in Britain*

Jens Lachmund, *Greening Berlin: The Co-Production of Science, Politics, and Urban Nature*

Chikako Takeshita, *The Global Biopolitics of the IUD: How Science Constructs Contraceptive Users and Women's Bodies*

Cyrus C. M. Mody, *Instrumental Community: Probe Microscopy and the Path to Nanotechnology*

Morana Alač, *Handling Digital Brains: A Laboratory Study of Multimodal Semiotic Interaction in the Age of Computers*

Gabrielle Hecht, editor, *Entangled Geographies: Empire and Technopolitics in the Global Cold War*

Michael E. Gorman, editor, *Trading Zones and Interactional Expertise: Creating New Kinds of Collaboration*

Matthias Gross, *Ignorance and Surprise: Science, Society, and Ecological Design*

Andrew Feenberg, *Between Reason and Experience: Essays in Technology and Modernity*

Wiebe E. Bijker, Roland Bal, and Ruud Hendricks, *The Paradox of Scientific Authority: The Role of Scientific Advice in Democracies*

Park Doing, *Velvet Revolution at the Synchrotron: Biology, Physics, and Change in Science*

Gabrielle Hecht, *The Radiance of France: Nuclear Power and National Identity after World War II*

Richard Rottenburg, *Far-Fetched Facts: A Parable of Development Aid*

Michel Callon, Pierre Lascoumes, and Yannick Barthe, *Acting in an Uncertain World: An Essay on Technical Democracy*

Ruth Oldenziel and Karin Zachmann, editors, *Cold War Kitchen: Americanization, Technology, and European Users*

Deborah G. Johnson and Jameson W. Wetmore, editors, *Technology and Society: Building Our Sociotechnical Future*

Trevor Pinch and Richard Swedberg, editors, *Living in a Material World: Economic Sociology Meets Science and Technology Studies*

Christopher R. Henke, *Cultivating Science, Harvesting Power: Science and Industrial Agriculture in California*

Helga Nowotny, *Insatiable Curiosity: Innovation in a Fragile Future*

Karin Bijsterveld, *Mechanical Sound: Technology, Culture, and Public Problems of Noise in the Twentieth Century*

Peter D. Norton, *Fighting Traffic: The Dawn of the Motor Age in the American City*

Joshua M. Greenberg, *From Betamax to Blockbuster: Video Stores tand the Invention of Movies on Video*

Mikael Hård and Thomas J. Misa, editors, *Urban Machinery: Inside Modern European Cities*

Christine Hine, *Systematics as Cyberscience: Computers, Change, and Continuity in Science*

Wesley Shrum, Joel Genuth, and Ivan Chompalov, *Structures of Scientific Collaboration*

Shobita Parthasarathy, *Building Genetic Medicine: Breast Cancer, Technology, and the Comparative Politics of Health Care*

Kristen Haring, *Ham Radio's Technical Culture*

Atsushi Akera, *Calculating a Natural World: Scientists, Engineers and Computers during the Rise of U.S. Cold War Research*

Donald MacKenzie, *An Engine, Not a Camera: How Financial Models Shape Markets*

Geoffrey C. Bowker, *Memory Practices in the Sciences*

Christophe Lécuyer, *Making Silicon Valley: Innovation and the Growth of High Tech, 1930–1970*

Anique Hommels, *Unbuilding Cities: Obduracy in Urban Sociotechnical Change*

David Kaiser, editor, *Pedagogy and the Practice of Science: Historical and Contemporary Perspectives*

Charis Thompson, *Making Parents: The Ontological Choreography of Reproductive Technology*

Pablo J. Boczkowski, *Digitizing the News: Innovation in Online Newspapers*

Dominique Vinck, editor, *Everyday Engineering: An Ethnography of Design and Innovation*

Nelly Oudshoorn and Trevor Pinch, editors, *How Users Matter: The Co-Construction of Users and Technology*

Peter Keating and Alberto Cambrosio, *Biomedical Platforms: Realigning the Normal and the Pathological in Late-Twentieth-Century Medicine*

Paul Rosen, *Framing Production: Technology, Culture, and Change in the British Bicycle Industry*

Maggie Mort, *Building the Trident Network: A Study of the Enrollment of People, Knowledge, and Machines*

Donald MacKenzie, *Mechanizing Proof: Computing, Risk, and Trust*

Geoffrey C. Bowker and Susan Leigh Star, *Sorting Things Out: Classification and Its Consequences*

Charles Bazerman, *The Languages of Edison's Light*

Janet Abbate, *Inventing the Internet*

Herbert Gottweis, *Governing Molecules: The Discursive Politics of Genetic Engineering in Europe and the United States*

Kathryn Henderson, *On Line and On Paper: Visual Representation, Visual Culture, and Computer Graphics in Design Engineering*

Susanne K. Schmidt and Raymund Werle, *Coordinating Technology: Studies in the International Standardization of Telecommunications*

Marc Berg, *Rationalizing Medical Work: Decision Support Techniques and Medical Practices*

Eda Kranakis, *Constructing a Bridge: An Exploration of Engineering Culture, Design, and Research in Nineteenth-Century France and America*

Paul N. Edwards, *The Closed World: Computers and the Politics of Discourse in Cold War America*

Donald MacKenzie, *Knowing Machines: Essays on Technical Change*

Wiebe E. Bijker, *Of Bicycles, Bakelites, and Bulbs: Toward a Theory of Sociotechnical Change*

Louis L. Bucciarelli, *Designing Engineers*

Geoffrey C. Bowker, *Science on the Run: Information Management and Industrial Geophysics at Schlumberger, 1920–1940*

Wiebe E. Bijker and John Law, editors, *Shaping Technology/Building Society: Studies in Sociotechnical Change*

Stuart Blume, *Insight and Industry: On the Dynamics of Technological Change in Medicine*

Donald MacKenzie, *Inventing Accuracy: A Historical Sociology of Nuclear Missile Guidance*

Pamela E. Mack, *Viewing the Earth: The Social Construction of the Landsat Satellite System*

H. M. Collins, *Artificial Experts: Social Knowledge and Intelligent Machines*

Lukas Engelmann and Christos Lynteris, *Sulphuric Utopias: A History of Maritime Fumigation*

http://mitpress.mit.edu/books/series/inside-technology

Printed in the United States
by Baker & Taylor Publisher Services